Structure and Biophysics – New Technologies for Current Challenges in Biology and Beyond

T0137684

NATO Science Series

A Series presenting the results of scientific meetings supported under the NATO Science Programme.

The Series is published by IOS Press, Amsterdam, and Springer in conjunction with the NATO Public Diplomacy Division.

Sub-Series

I. Life and Behavioural Sciences	IOS Press
II. Mathematics, Physics and Chemistry	Springer
III. Computer and Systems Science	IOS Press
IV. Earth and Environmental Sciences	Springer

The NATO Science Series continues the series of books published formerly as the NATO ASI Series.

The NATO Science Programme offers support for collaboration in civil science between scientists of countries of the Euro-Atlantic Partnership Council. The types of scientific meeting generally supported are "Advanced Study Institutes" and "Advanced Research Workshops", and the NATO Science Series collects together the results of these meetings. The meetings are co-organized by scientists from NATO countries and scientists from NATO's Partner countries — countries of the CIS and Central and Eastern Europe.

Advanced Study Institutes are high-level tutorial courses offering in-depth study of latest advances in a field.
Advanced Research Workshops are expert meetings aimed at critical assessment of a field, and identification of directions for future action.

As a consequence of the restructuring of the NATO Science Programme in 1999, the NATO Science Series was re-organized to the four sub-series noted above. Please consult the following web sites for information on previous volumes published in the Series.

http://www.nato.int/science
http://www.springer.com
http://www.iospress.nl

Series II: Mathematics, Physics and Chemistry – Vol. 231

Structure and Biophysics – New Technologies for Current Challenges in Biology Beyond

edited by

Joseph D. Puglisi

Stanford University,
Stanford, California, U.S.A.

Springer

Published in cooperation with NATO Public Diplomacy Division

Proceedings of the NATO Advanced Study Institute on
Structure and Biophysics – New Technologies for Current Challenges
in Biology and Beyond
Erice, Italy
22 June–3 July 2005

A C.I.P. Catalogue record for this book is available from the Library of Congress.

ISBN 978-1-4020-5899-8 (PB)
ISBN 978-1-4020-5900-1 (eBook)

Published by Springer,
P.O. Box 17, 3300 AA Dordrecht, The Netherlands.

www.springer.com

Printed on acid-free paper

PREFACE

This volume is a collection of articles from the Proceedings of the International School of Structural Biology and Magnetic Resonance 7th Course: Structure, Structure and Biophysics – New Technologies for Current Challenges in Biology and Beyond.

This NATO Advance Study Institute (ASI) was held in Erice (Italy) at the Ettore Majorana Foundation and Centre for Scientific Culture on 22 June through 3 July 2005. The ASI brought together a diverse group of experts in the fields of structural biology, biophysics, and physics. Prominent lecturers, from seven different countries, and students from around the world participated in the NATO ASI organized by Professors Joseph Puglisi (Stanford University, California, USA) and Alexander Arseniev (Moscow, RU). Advances in nuclear magnetic resonance (NMR) spectroscopy and X-ray crystallography have allowed the three-dimensional structures of many biological macromolecules and their complexes, including the ribosome and RNA polymerase to be solved. Fundamental principles of NMR spectroscopy and dynamics, X-ray crystallography, computation, and experimental dynamics were taught in the context of important biological applications. The ASI addressed the treatment and detection of bioterrorism agents, and focused on critical partner country priorities in biotechnology, materials, and drug discovery.

The range of topics here represents the diversity of critical problems between structural biology, biochemistry, and biophysics, in which lies the fertile ground of drug development, biotechnology, and new materials. The individual articles represent the state of the art in each area and provide a guide to the original literature in this rapidly developing field.

CONTENTS

STRUCTURAL GENOMICS BY NMR SPECTROSCOPY

ALEKSANDRAS GUTMANAS[1,2,3], ADELINDA YEE[1,3], SAMPATH
SRISAILAM[1,2,3], AND CHERYL H. ARROWSMITH[1,2*,3]

[1]*Ontario Centre for Structural Proteomics, University Health Network, Toronto,
Ontario, Canada*
[2]*Department of Medical Biophysics, University of Toronto, Toronto, Canada*
[3]*Northeast Structural Genomics Consortium*
Corresponding author e-mail: carrow@uhnres.utoronto.ca
Address: 100 College St, Room 351C, Banting Building, Tornto, Ontario, Canada
Tel: +1 416 946 0881
Fax: +1 416 946 0880

Abstract: Structural genomics is becoming an established approach for protein structure determination to more quickly annotate newly sequenced genomes with structural information. The simultaneous use of complementary tools of X-ray crystallography and nuclear magnetic resonance (NMR) enhance its overall performance. This paper presents the NMR pipeline from target selection through structure determination and uses structural proteomics in Toronto (SPiT) as a successful example of such an approach. During the first 5 years of the group's existence 48 novel NMR structures (less that 30% sequence identity to proteins with known structures) were solved and deposited in the protein data bank (PDB). Efforts are being made to optimize sample conditions, reduce the required measurement time, and partially automate data analysis, resonance and nuclear overhauser effect (NOE) assignment, and structure determination.

1. Introduction

In recent years the successful sequencing of many organisms' DNA, the Human Genome Project being the most prominent such undertaking, raised a variety of new questions with regards to the newly available data. Valuable as this gene and protein sequence information is in its own right for the diverse studies in biochemistry, evolutionary biology and molecular biology, it becomes increasingly important to understand the structural and ultimately functional characteristics of the multitude of proteins acting in our or other species' bodies (Christendat et al., 2000; Montelione et al., 2000; Yee et al., 2003).

In contrast to traditional structural biology approach, in which the quest for protein structure commences subsequent to extensive biochemical investigations, the structural genomics approach aims at providing the scientific community with as many and as divergent protein structures as possible. Ever improving bioinformatics and modeling tools allow one to routinely predict structures or at the very least folds of homologous proteins with at least 30% sequence identity. However, modeling is less reliable for lower degrees of sequence identity, which calls for employment of experimental techniques. The two mainstream methods for solving protein structures are NMR and X-ray crystallography. The latter is more flexible in terms of the protein sizes and its data analysis is fairly straightforward once high-resolution

J. D. Puglisi (ed.), Structure and Biophysics – New Technologies for Current Challenges in Biology and Beyond, 1–11.
© *2007 Springer.*

diffraction data is obtained. However, its major drawback is the crystallization process, which is still insufficiently understood and whose outcome is oftentimes unpredictable. Solution NMR does not need crystals but instead requires stable concentrated (~1 mM) samples of isotope-labeled proteins. Its drawbacks lie in the perceived size limit of the studied macromolecules (currently at 20–30 kDa for a "straightforward" structure determination) as well as long data collection times (~4 weeks per sample) and the intrinsic complexity of the data analysis, which may take up to a few months depending on the sample. These limitations are being constantly addressed by new experimental hardware such as high-field spectrometers and cryogenic probes, providing higher resolution and sensitivity, by novel experiments enabling a more rapid and efficient data collection (Kim and Szyperski, 2003; Luan et al., 2005; Kupče and Freeman, 2004; Rovnyak et al., 2004), and by automation efforts (Malmodin et al., 2003; Zimmerman et al., 1997), slashing the time required for data analysis and structure calculation. These advances become especially welcome in the context of structural proteomics by NMR (Yee et al., 2002), as they can substantially reduce the need for the expensive spectrometer time and facilitate the human effort and thus ultimately yield more structures per invested dollar.

It is noteworthy that the two experimental techniques, NMR and X-ray crystallography, are complementary. Only 8% of protein targets are amenable to structure determination by both methods, as was shown by a study of 263 unique protein sequences screened for both NMR and X-ray crystallography (Yee et al., 2005). Similar conclusions were reached by other groups (Snyder et al., 2005; Tyler et al., 2005). In at least one other structural genomics group, however, NMR is used solely as a prescreening technique for the X-ray crystallography (Page et al., 2005). This paper mainly concentrates on the application of NMR in structural proteomics and uses our group of "structural proteomics in Toronto" (SPiT) as a successful example of such an application (http://www.uhnresearch.ca/centres/proteomics). We are affiliated with the Northeast Structural Genomics Consortium, part of the U.S. National Institutes of Health protein structure initiative (PSI, http://www.nesg.org).

Currently, SPiT targets soluble proteins from a number of species, mainly microorganisms. Since 1999, when the work commenced, 188 total structures have been solved; 48 of these by NMR spectroscopy. All structures are deposited in the protein data bank (PDB) without delay, ensuring rapid and ready access to these results by the scientific community.

2.   Results and Discussion

2.1. ORGANIZATIONAL SETUP

The NMR structural proteomics group in Toronto is currently composed of ten people, four of whom are directly involved in protein production; two are involved in structure determination using the currently accepted conventional methods; one is mainly responsible for data acquisition and the rest work on NMR methods development. The cloning for the NMR targets is centralized with those for crystallography, likewise management of the enormous amount of data generated by the "pipeline" is centralized under a common database.

2.2. THE PIPELINE

The NMR pipeline has evolved considerably since it was first initiated in Toronto. The diagram in Figure 1 is a representation of the currently implemented pipeline. The first half shows the

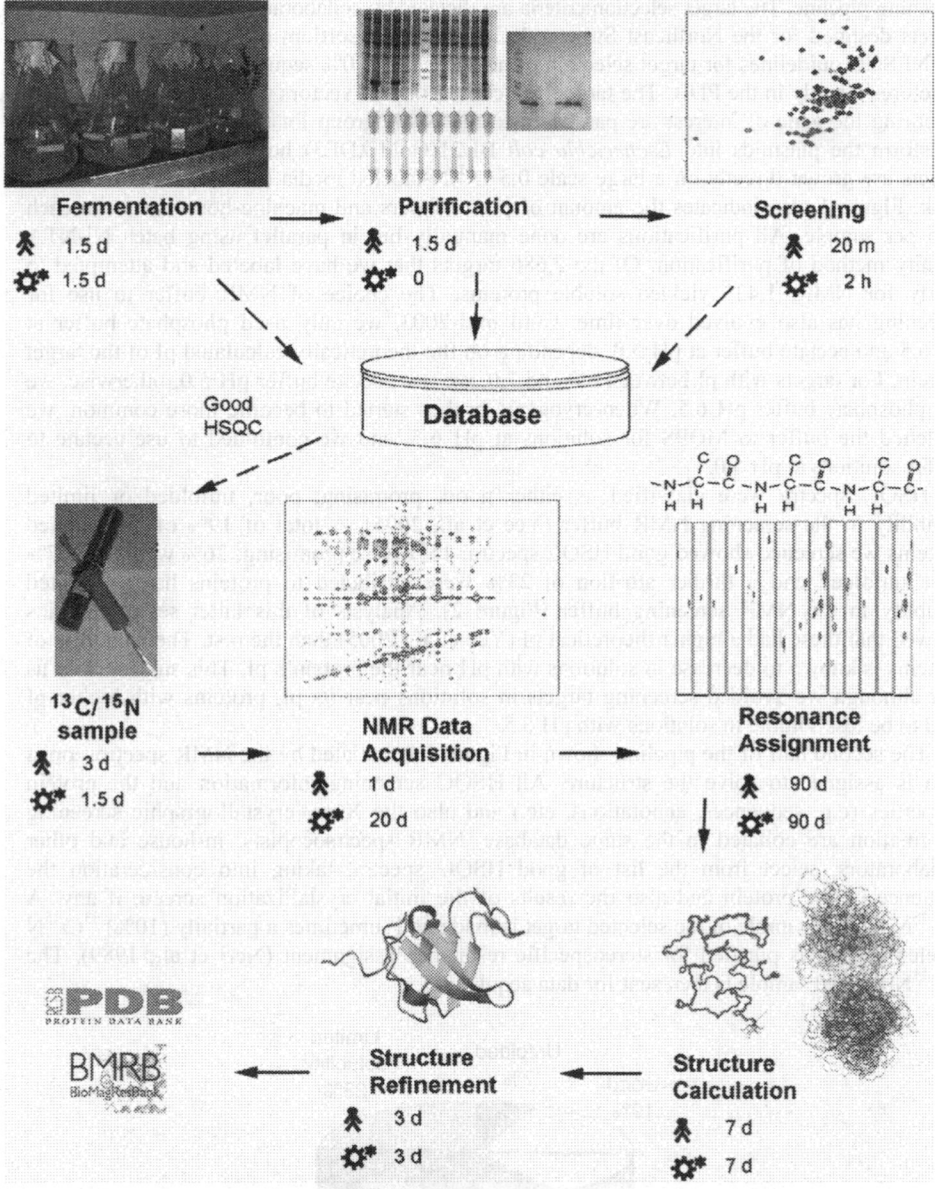

Figure 1. Representation of the currently implemented structural proteomics pipeline by NMR at the structural proteomics in Toronto (SPiT). *Dashed arrows* show the flow of information into and out of the common database. The average person-hour and machine-hour indicated is on a per sample basis.

screening pipeline. The target selection criteria are dictated by collaborating groups, for example, targets destined for the Northeast Structural Genomics Consortium (NESGC) would follow the NESGC guidelines for target selection of no more than 30% sequence identity with those structures already in the PDB. The targets are cloned in pET vectors (Novagen) and plasmids harboring the gene of interest are passed on to the NMR group for screening. We currently transform the plasmids into *Escherichia coli* BL21(Gold λDE3) host expression cells. All targets are grown directly on a large scale 0.5 L ^{15}N-labeled media (2XM9) in a 4L baffled flask. Figure 1 also indicates the amount of person-hours and machine-hours spent on each step per sample. All purifications are done manually but in parallel using batch Ni-NTA affinity method of purification. Of the 2,686 targets that we have labeled and attempted to purify for NMR, 1,433 yielded soluble proteins. The choice of NMR buffer to use for screening has also evolved over time. Until mid-2003, we only used phosphate buffer at pH 6.5 and acetate buffer at pH 5.0, depending on the theoretically calculated pI of the target protein. For targets with pI between 6.0 and 7.0, we use acetate buffer pH 5.0, otherwise, we use phosphate buffer pH 6.5. When cryogenic probes started to become more common, we switched the buffer to MOPS for solutions at pH 6.5, and we continued to use acetate to buffer solutions at pH 5.0.

HSQC spectra were classified as either good, promising, poor, unfolded or limited solubility in the screening NMR buffer (Yee et al., 2005). A total of 19% of the purified proteins we screened showed good HSQC spectra; 15% were promising; 36% were poor; 7% were unfolded and a further attrition of 23% were attributed to proteins having limited solublity in the NMR screening buffer (Figure 2). Analysis of this latter set of proteins showed that these had a higher theoretical pI (Yee et al., 2005) than the rest. The solubility of proteins is known to decrease in solutions with pH near the protein's pI. This suggested to us that although we avoided screening targets in solutions near its pI, proteins with higher pI tend to be less soluble in solutions with pH 6.5.

The second half of the pipeline shown in Figure 1 is initiated by the NMR spectroscopist who is assigned to solve the structure. All HSQC screening information and the protein properties (e.g. sequences, annotations, etc.) and also the X-ray crystallographic screening information are collated in the same database. NMR spectrsocopists, in-house and other collaborators, select from the list of good HSQC spectra, taking into consideration the sequence of the protein and also the results of the initial crystallization screen, if any. A ^{13}C/^{15}N-labeled sample of the selected target is made and sometimes a partially (10%) ^{13}C/^{15}N labeled protein is purified for stereospecific resonance assignment (Neri et al., 1989). The ^{13}C/^{15}N-labeled sample is then sent for data acquisition.

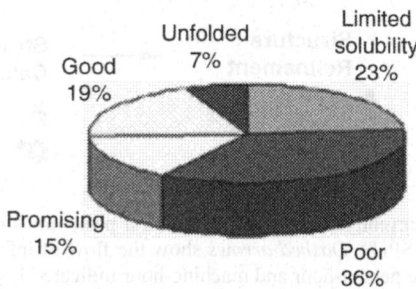

Figure 2. Distribution of the HSQC classifications of purified proteins as of November 2005.

2.3. STRUCTURE DETERMINATION

The exact combination of the recorded NMR experiments varies between participating groups, yet normally conventional triple resonance and side chain techniques are used for resonance assignment, while distance and dihedral angle constraints are derived from nuclear overhauser effect (NOE) spectra and measurement of coupling constants. These constraints are then used to calculate the structure ensemble by one or more of the available programs for structure calculation, such as CYANA (Güntert, 2004), Autostructure (Zheng et al., 2003) and CNS/ARIA (Linge et al., 2003). A number of new experiments are currently being developed and implemented to speed up the data acquisition, notable examples being G-matrix Fourier transform (GFT) (Kim and Szyperski, 2003) and related techniques and nonuniform sampling methods such as multidimensional decomposition (Luan et al., 2005). Another prospect for simplification of the structure determination process is to minimize the necessity for manual resonance and NOE assignments using such programs as Autoassign and ABACUS (Grishaev et al., 2005). These and other novel methods are reviewed below in the *New Developments* section.

Table 1 summarizes the current progress for each genome. So far, out of 228 targets yielding good HSQC spectra, 48 NMR structures from a variety of genomes have been deposited in

TABLE 1. Progress of structure solution by NMR for each attempted SPiT genome target

Organism[a]	Screened[b]	Good[c]	Solved[d]	In progress[e]	Work stopped[f]
Agrobacterium tumefaciens C58 (ATC)	83	13	0	2	0
Archaeoglobus fulgidus (AF)	20	3	0	2	0
Bacillus subtilis (BSU)	1	0	0	0	0
Deinococcus radiodurans (DRD)	18	2	0	0	0
Encephalitozoon cuniculi (ECU)	79	9	0	3	2
Escherichia coli K12 (EC)	147	51	7	14	16
Helicobacter pylori (HP)	7	0	0	0	0
Homo sapiens (HS)	19	5	1	0	2
Methanothermobacter thermautotrophicus (MTH)	229	42	22	10	6
Nitrosomonas europea (NE)	104	12	1	5	3
Pseudomonas aeruginosa (PA)	150	25	3	7	1
Rhodopseudomonas palustris CGA009 (RP)	168	22	1	0	0
Saccaromyces cerivisiae (YST)	223	30	5	2	10
Sulfolobus solfataricus (SSO)	20	4	1	1	1
Thermoplasma acidophilum (TA)	176	40	1	12	6
Thermotoga maritima (TM)	82	25	3	2	15
Bacteriophage HK97 (HK97GP)	15	2	0	0	0
Bacteriophage lambda (LM)	14	2	2	0	0
Myxoma virus (MYX)	11	1	1	0	0
Total[g]	1566	288	48	60	62

[a]Organism from which the target is cloned.
[b]Total number of screened samples for each organism. This number includes good, promising and poor HSQC spectra as well as unfolded and samples with limited solubility in the NMR screening buffer.
[c]Number of HSQC spectra judged to be good and amenable to strcture solution by NMR.
[d]Number of structures solved as part of SPiT.
[e]Number of targets sent to collaborators or reserved for current and future work.
[f]Number of structures or orthologs solved by other groups.
[g]Totals for all genomes.

the PDB (www.rcsb.org/pdb). Another 60 targets are in progress in-house and in other collaborating laboratories or already reserved for later analysis. Additional 62 targets were abandoned at various stages because structures of identical or homologous (with more than 30% sequence identity) proteins were deposited by other groups. A small percentage of targets were abandoned during the data analysis because they exhibited multiple conformations not readily detectable in the initial HSQC screen. Finally, a number of proteins are still available for solution either by SPiT NMR group or by collaborators (http://www. uhnresearch. ca/ centres/proteomics).

From Figure 3, which presents the percentages of HSQC screens that yielded good spectra for each genome, it emerges that number of good HSQC spectra varies roughly between 10% and 30% of total number of screened samples with *E. coli* having a maximum 34%. Three thermophilic organisms (*Thermotoga maritima*, TM, *T. acidophilum*, TA, and *Methanothermobacter thermautotrophicus*, MTH) appear to yield good HSQC spectra more readily than others, however, the present comparison is biased. For several mesophilic genomes that were started later in our project (e.g. *Pseudomonas aeruginosa*, PA) a fair number of structures were solved by other groups and therefore we did not attempt to screen those proteins. Had we included these genes in the analysis, the "success rate" for these genomes may

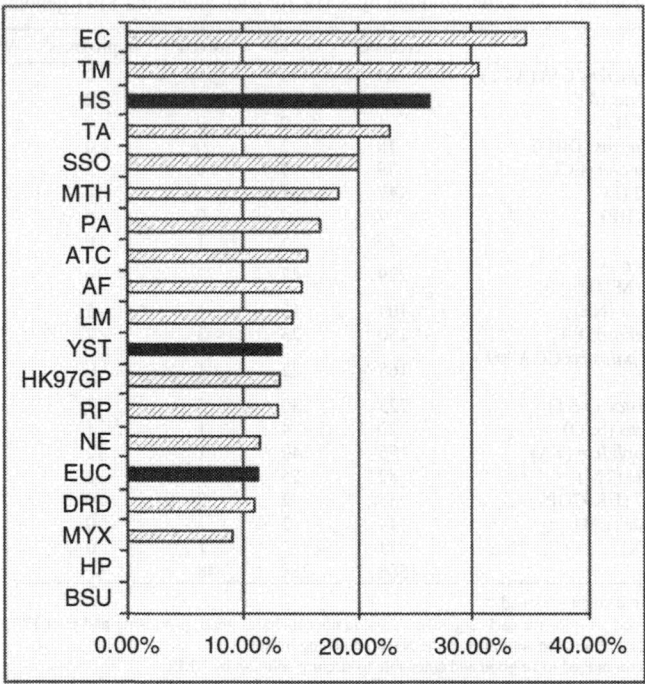

Figure 3. Percentages of HSQC screens for all attempted genomes producing good spectra and indicating that the corresponding targets can proceed to structure determination. Eukaryotic organisms are indicated by *solid bars*. The abbreviations for organism names explained in Table 1.

be higher. A thorough comparison between *T. maritima* and *E. coli* was conducted earlier and no tendency was observed concerning quality of HSQC spectra for the two species (Savchenko et al., 2003). Certainly, optimizing constructs and changing buffer conditions would aid in increasing solubility and finding suitable cofactors could increase these percentages. However, as Table 1 also indicates, the major bottleneck in the current arrangement is still NMR data collection and interpretation as at least a few dozen proteins having passed screening stage are on hold waiting for an NMR spectroscopist to solve them. Limited spectrometer time is a particularly impeding factor in this latter stage and a number of techniques described below (e.g. GFT, Kim and Szyperski, 2003; MDD, Luan et al., 2005) address this issue. Data interpretation is still a highly interactive process, which requires human judgment in deciding ambiguous situations. As the diagram on Figure 1 indicates, it can take up to 3 months for resonance and NOE assignments, while it is a matter of 1 or 2 weeks to obtain a sample and screen it. Work on automation in interpreting the spectra is also discussed in a later section.

3. New Developments in NMR Applicable to Structural Proteomics

During the past few years a number of developments in NMR methodology permit one to both reduce the time and effort required for data collection and analysis. Also, novel experiments are designed, which allow one to probe different physical properties of proteins. A few of these techniques and their implications for structural proteomics are surveyed in this section.

3.1. CRYOGENIC PROBES AND HIGH-FIELD SPECTROMETERS

Advances in instrumentation are certainly one of the major areas of improvement in NMR. In last 4 or 5 years, 800 MHz and even 900 MHz spectrometers became much more stable and are now employed on a regular basis for the purposes of structural proteomics. They improve the resolution and sensitivity of the experiments and are becoming indispensable in NMR, in particular for studies of larger macromolecules (>30 kDa). Although this class of proteins is not currently targeted by the NMR branch of structural proteomics, it is a matter of time before it is. In parallel with increasing the field strength, another line of instrumentation improvement during this period was the creation and implementation of cryogenically cooled probes, which greatly (up to 5 times) reduce the electronic noise in comparison to the room temperature counterparts.

In combination with the high magnetic field of modern spectrometers, the cryogenic probes allow reduction of measurement time for most samples, thus relaxing the requirement for the sample stability. A different benefit of these technologies is the ability to use less concentrated samples (<1 mM) and maintain good signal to noise (S/N) ratio. Improvements to instrumentation have certainly not reached their peak and future developments will bring new advances in applicability of NMR.

3.2. DEVELOPMENTS IN EXPERIMENTAL TECHNIQUES

The advances in instrumentation have opened new avenues to the design of pulse sequences and probing previously unused physical properties of proteins. Among such experiments is Transverse relaxation optimization spectroscopy (TROSY; Pervushin et al., 1997), which is particularly useful for large and/or membrane proteins as it circumvents the line broadening associated with slow tumbling motion of large molecules. The group of proteins currently targeted by structural proteomics by NMR (<25 kDa), however, benefits only moderately from this technique.

In many instances, and in particular for a number of triple resonance experiments with limited number of peaks in the spectrum, collection of a complete data set may seem superfluous, since the data could have an order of magnitude lower S/N and still be unambiguously interpreted. Several techniques are being developed to take advantage of this fact. The common feature is that they sample only a subset of the underlying N-dimensional spectrum. One set of these techniques, namely reduced dimensionality (RD) (Szyperski et al., 2002), and G-matrix Fourier Transform, or GFT (Kim and Szyperski, 2003), use such subsets to directly obtain the relevant parameters of the N-dimensional spectrum, such as peak positions, from the recorded spectra with lower dimensionality, thereby circumventing the reconstruction of a complete underlying spectrum. Thus, new strategies in interpreting the data are necessary. Other approaches, such as projection reconstruction (PR) methods (Kupče and Freeman, 2004), use a very similar strategy in choosing the subset to record, yet their ultimate aim is to reproduce the complete underlying spectrum, which would then be analyzed in a more traditional way. It is also possible to interpret the projection planes individually and generate the peak list of the underlying spectrum as recently reported by Hiller et al. (2005).

Still other methods, in particular maximum entropy reconstruction (Rovnyak et al., 2004) and multidimensional decomposition (Luan et al., 2005) are more conservative in that the recorded subsets are usually larger than for the techniques mentioned above. The subsets are also selected randomly with an optional bias for shorter delays. As a collective set these methods are certainly welcome in structural proteomics as they can significantly (between 3 and 10 and even up to 100 times) reduce the necessary experimental time, allowing for higher throughput of samples. So far only multidimensional decomposition has been reliably applied to experimental NOESY data and it has consistently reproduced full dynamic range of intensities (Luan et al., 2005), while other methods find their applicability in triple-resonance spectra. These developments make collection of higher dimensional spectra (4D and above) feasible on a routine basis.

3.3. RESIDUAL DIPOLAR COUPLINGS

Although the experiments for measurements of one-bond dipolar couplings have been used for some time, it is only recently that they became more routine. The sample is placed in an anisotropic medium, such as a bicelle, phage or gel, which slightly orients the protein and hence the coupling of the N–H or C–H bond is no longer averaged to zero. The information extracted from the residual dipolar couplings is useful in determining the relative domain orientation, validation of independently determined structures or for refinement of NMR structures (Prestegard et al., 2004).

3.4. AUTOMATION OF DATA ANALYSIS

The major bottleneck of the structure determination by NMR is the complicated and tedious process of data interpretation. Normally, the resonances of the backbone nuclei are assigned and mapped to the sequence from a set of triple resonance experiments. Side chain nuclei are then assigned from a different set of triple resonance spectra. Next the peaks of the NOE spectra are assigned to specific protons in the protein and finally distance constraints are derived from NOE peaks. A number of programs in the field greatly assist in the process, such as Autoassign (Zimmerman et al., 1997), MONTE (Hitchens et al., 2003) and GARANT (O'Connell et al., 1999). Peak picking, which is one of the most unwelcome tasks of NMR spectroscopist, can be tackled by such software as AUTOPSY (Koradi et al., 1998) and MUNIN (Gutmanas et al., 2002).

A different approach is taken by the program ABACUS (Grishaev et al., 2005), which uses fewer spectra for the same purpose and attempts to collect resonances into spin systems without actually mapping them onto the amino acid sequence until NOE data is introduced. This eliminates the necessity to assign the resonances and in particular the NOE peaks before deriving the distance constraints, which is a major source of ambiguity. So far ABACUS showed promising performance.

4. Conclusions

In conclusion, NMR is a viable alternative for protein structure determination in the framework of structural genomics and can be a valuable complement to X-ray crystallography. The NMR pipeline is only in the beginning of its development, yet it has already produced 48 new structures originating from our group in SPiT. A number of technological and computational innovations can accelerate this process. Several experimental techniques, such as GFT, projection-reconstruction, nonuniform acquisition, in combination with software for multidimensional decomposition or maximum entropy reconstruction, are being developed or tested and in a few cases employed by our group. Automatic assignment tools, such as Autoassign and ABACUS also become indispensable for rapid and accurate interpretation of NMR data.

References

1. Christendat, D., Yee, A., Dharamsi, A., Kluger, Y., Savchenko, A., Cort, J.R., Booth, V., Mackereth, C.D., Saridakis, V., Ekiel, I., Kozlov, G., Maxwell, K.L., Wu, N., McIntosh, L.P., Gehring, K., Kennedy, M.A., Davidson, A.R., Pai, E.F., Gerstein, M., Edwards, A.M., and Arrowsmith, C.H. (2000). Structural proteomics of an archaeon. *Nat. Struct. Biol.* 7, 903–909.
2. Grishaev, A., Steren, C.A., Wu, B., Pineda-Lucena, A., Arrowsmith, C., and Llinas, M. (2005). ABACUS, a direct method for protein NMR structure computation via assembly of fragments. *Proteins* 61, 36–43.
3. Güntert, P. (2004). Automated NMR structure calculation with CYANA. *Methods Mol. Biol.* 278, 353–378.
4. Gutmanas, A., Jarvoll, P., Orekhov, V.Y., and Billeter, M. (2002). Three-way decomposition of a complete 3D 15N-NOESY-HSQC. *J. Biomol. NMR* 24, 191–201.

5. Hiller, S., Fiorito, F., Wüthrich, K., and Wider, G. (2005). Automated projection spectroscopy (APSY). *Proc. Natl. Acad. Sci. USA* 102, 10876–10881.

6. Hitchens, T.K., Lukin, J.A., Zhan, Y., McCallum, S.A., and Rule, G.S. (2003). MONTE: an automated Monte Carlo based approach to nuclear magnetic resonance assignment of proteins. *J. Biomol. NMR* 25, 1–9.

7. Kim, S. and Szyperski, T. (2003). GFT NMR, a new approach to rapidly obtain precise high-dimensional NMR spectral information. *J. Am. Chem. Soc.* 125, 1385–1393.

8. Koradi, R., Billeter, M., Engeli, M., Guntert, P., and Wuthrich, K. (1998). Automated peak picking and peak integration in macromolecular NMR spectra using AUTOPSY. *J. Magn. Reson.* 135, 288–297.

9. Kupče, E. and Freeman, R. (2004). Projection-reconstruction technique for speeding up multidimensional NMR spectroscopy. *J. Am. Chem. Soc.* 126, 6429–6440.

10. Linge, J.P., Habeck, M., Rieping, W., and Nilges, M. (2003). ARIA: automated NOE assignment and NMR structure calculation. *Bioinformatics* 19, 315–316.

11. Luan, T., Jaravine, V., Yee, A., Arrowsmith, C.H., and Orekhov, V.Y. (2005). Optimization of resolution and sensitivity of 4D NOESY using multi-dimensional decomposition. *J. Biomol. NMR* 33, 1–14.

12. Malmodin, D., Papavoine, C.H., and Billeter, M. (2003). Fully automated sequence-specific resonance assignments of hetero-nuclear protein spectra. *J. Biomol. NMR* 27, 69–79.

13. Montelione, G.T., Zheng, D., Huang, Y., and Szyperski, T. (2000). Protein NMR spectroscopy in structural genomics. *Nat. Struct. Biol.* 7, 982–984.

14. Neri, D., Szyperski, T., Otting, G., Senn, H., and Wüthrich, K. (1989). Stereospecific nuclear magnetic resonance assignments of the methyl groups of valine and leucine in the DNA-binding domain of the 434 repressor by biosynthetically directed fractional 13C labeling. *Biochemistry* 28, 7510–7516.

15. O'Connell, J.F., Pryor, K.D., Grant, S.K., and Leiting, B. (1999). A high quality nuclear magnetic resonance solution structure of peptide deformylase from *Escherichia coli*: application of an automated assignment strategy using GARANT. *J. Biomol. NMR* 13, 311–324.

16. Page, R., Peti, W., Wilson, I.A., Stevens, R.C., and Wüthrich, K. (2005). NMR screening and crystal quality of bacterially expressed prokaryotic and eukaryotic proteins in a structural genomics pipeline. *Proc. Natl. Acad. Sci. USA* 102, 1901–1905.

17. Pervushin, K., Riek, R., Wider, G., and Wuthrich, K. (1997). Attenuated T2 relaxation by mutual cancellation of dipole-dipole coupling and chemical shift anisotropy indicates an avenue to NMR structures of very large biological macromolecules in solution. *Proc. Natl. Acad. Sci. USA* 94, 12366–12371.

18. Prestegard, J.H., Bougault, C.M., and Kishore, A.I. (2004). Residual dipolar couplings in structure determination of biomolecules. *Chem. Rev.* 104, 3519–3540.

19. Rovnyak, D., Frueh, D.P., Sastry, M., Sun, Z.Y., Stern, A.S., Hoch, J.C., and Wagner, G. (2004). Accelerated acquisition of high resolution triple-resonance spectra using non-uniform sampling and maximum entropy reconstruction. *J. Magn. Reson.* 170, 15–21.

20. Savchenko, A., Yee, A., Khachatryan, A., Skarina, T., Evdokimova, E., Pavlova, M., Semesi, A., Northey, J., Beasley, S., Lan, N., Das, R., Gerstein, M., Arrowmith, C.H., and Edwards, A.M. (2003). Strategies for structural proteomics of prokaryotes: Quantifying the advantages of studying orthologous proteins and of using both NMR and X-ray crystallography approaches. *Proteins* 50, 392–399.

21. Snyder, D.A., Chen, Y., Denissova, N.G., Acton, T., Aramini, J.M., Ciano, M., Karlin, R., Liu, J., Manor, P., Rajan, P.A., Rossi, P., Swapna, G.V.T., Xiao, R., Rost, B., Hunt, J., and Montelione, G.T. (2005). Comparisons of NMR spectral quality and success in crystallization demonstrate that NMR and X-ray crystallography are complementary methods for small protein structure determination. *J. Am. Chem. Soc.* 127, 16505–16511.

22. Szyperski, T., Yeh, D.C., Sukumaran, D.K., Moseley, H.N., and Montelione, G.T. (2002). Reduced-dimensionality NMR spectroscopy for high-throughput protein resonance assignment. *Proc. Natl. Acad. Sci. USA* 99, 8009–8014.
23. Tyler, R.C., Aceti, D.J., Bingman, C.A., Cornilescu, C.C., Fox, B.G., Frederick, R.O., Jeon, W.B., Lee, M.S., Newman, C.S., Peterson, F.C., Phillips, G.N. Jr., Shahan, M.N., Singh, S., Song, J., Sreenath, H.K., Tyler, E.M., Ulrich, E.L., Vinarov, D.A., Vojtik, F.C., Volkman, B.F., Wrobel, R.L., Zhao, Q., and Markley, J.L. (2005). Comparison of cell-based and cell-free protocols for producing target proteins from the *Arabidopsis thaliana* genome for structural studies. *Proteins* 59, 633–643.
24. Yee, A., Chang, X., Pineda-Lucena, A., Wu, B., Semesi, A., Le, B., Ramelot, T., Lee, G.M., Bhattacharyya, S., Gutierrez, P., Denisov, A., Lee, C.H., Cort, J.R., Kozlov, G., Liao, J., Finak, G., Chen, L., Wishart, D., Lee, W., McIntosh, L.P., Gehring, K., Kennedy, M.A., Edwards, A.M., and Arrowsmith, C.H. (2002). An NMR approach to structural proteomics. *Proc. Natl. Acad. Sci. USA* 99, 1825–1830.
25. Yee, A., Pardee, K., Christendat, D., Savchenko, A., Edwards, A.M., and Arrowsmith, C.H. (2003). Structural proteomics: toward high-throughput structural biology as a tool in functional genomics. *Acc. Chem. Res.* 36, 183–189.
26. Yee, A.A., Savchenko, A., Ignachenko, A., Lukin, J., Xu, X., Skarina, T., Evdokimova, E., Liu, C.S., Semesi, A., Guido, V., Edwards, A.M., and Arrowsmith, C.H. (2005). NMR and x-ray crystallography, complementary tools in structural proteomics of small proteins. *J. Am. Chem. Soc.* 127, 16512–16517.
27. Zheng, D., Huang, Y.J., Moseley, H.N., Xiao, R., Aramini, J., Swapna, G.V., and Montelione, G.T. (2003). Automated protein fold determination using a minimal NMR constraint strategy. *Protein Sci* 12, 1232–1246.
28. Zimmerman, D.E., Kulikowski, C.A., Huang, Y., Feng, W., Tashiro, M., Shimotakahara, S., Chien, C., Powers, R., and Montelione, G.T. (1997). Automated analysis of protein NMR assignments using methods from artificial intelligence. *J. Mol. Biol.* 269, 592–610.

INVESTIGATION OF PROTEINS IN LIVING BACTERIA WITH IN-CELL NMR EXPERIMENTS

VOLKER DÖTSCH
Institute of Biophysical Chemistry
University of Frankfurt
Max-Von-Laue Str. 9
60439 Frankfurt
Germany

1.   Introduction

Since its discovery in 1945 the nuclear magnetic resonance (NMR) effect has developed from an interesting physical phenomenon into the most important analytical technique in chemistry. Originally deemed to be too insensitive for applications to biological systems many different techniques all based on the NMR phenomenon have emerged during the last 30 years. Today NMR spectroscopy is used for investigations of structure and dynamics of macromolecules as well as binding studies. One aspect that distinguishes NMR from other biophysical techniques is that it can be applied not only to purified in vitro samples of defined composition but also to investigations of living cells as well as entire organisms. In particular imaging techniques have had and continue to have an enormous impact as a tool in medicine. Imaging techniques use differences in the properties of protons between different tissues to create images. Characteristics that can be used include differences in the T_1 relaxation rates, or in the rate of diffusion. Certain differences such as those in the relaxation rates can be enhanced by adding external agents such as gadolinium salts that are injected into the organism being investigated. In addition to protons other nuclei have also been used for imaging purposes, for example ^{13}C, ^{10}B, and ^{11}B.

While imaging techniques provide information about entire organs, NMR can also be used to obtain detailed information about molecular processes in living cells and organisms. In vivo NMR spectroscopy has for example been used to investigate metabolic fluxes (Bachert, 1998; Cohen et al., 1989; Li et al., 1996; Gillies, 1994; Kanamori and Ross, 1997). In these experiments small molecules labeled with NMR active isotopes, for example ^{13}C, are added to cells. Through the cellular metabolism these NMR active spins get incorporated into other molecules, which can be identified based on their characteristic chemical shifts.

In recent years NMR spectroscopy has also developed into an important tool for the investigation of body fluids such as urine (Bollard et al., 2005a, b). While these metabonomic investigations are, strictly speaking, not an in vivo application, they have an important and growing influence in the field of disease diagnosis. These fluids contain hundreds of small molecules that produce many NMR-detectable signals. While the identification of the individual components of these mixtures is too complicated, changes in the overall signal pattern can be characteristic for certain types of diseases.

Until recently, investigations of molecules in living cellular systems were limited to these classical in vivo (and metabonomics) NMR experiments that focus on small molecules. The gap between in vivo NMR experiments with small metabolic molecules on the one hand, and

J. D. Puglisi (ed.), Structure and Biophysics – New Technologies for Current Challenges in Biology and Beyond, 13–17.
© 2007 *Springer.*

imaging techniques providing information about entire organs and organisms on the other hand has been closed by the development of techniques that allow researchers to investigate the behavior of biological macromolecules inside living cells (Serber and Dötsch, 2001; Serber et al., 2001a, b, 2004a; Wieruszeski et al., 2001; Dedmon et al., 2002; Hubbard et al., 2003). The observation of biological macromolecules in cellular systems is based on labeling schemes that enable the selective identification of these macromolecules in an environment that is crowded with other macromolecules as well as with many different small molecules. Basically two different approaches to achieve this goal exist. The first is based on micro-injecting a highly concentrated sample of a purified macromolecule labeled with NMR active isotopes into the cell type of interest (Selenko and Wagner, 2004; Serber et al., 2004b). Other techniques that allow proteins to pass through the cellular membrane such as arginine tags or dendrite based systems might also be possible. The advantage of these techniques is that the natural abundance of the NMR active isotopes (1.1% ^{13}C and 0.2% ^{15}N) represents the only potential background signals. The main disadvantage of these techniques is that they are very labor-intensive and technically very demanding, and are therefore just emerging. The method used most often up to now is the expression of the macromolecule of interest inside cells, mainly in bacterial cells (Serber and Dötsch, 2001; Serber et al., 2001a, b, 2004a; Wieruszeski et al., 2001; Dedmon et al., 2002; Hubbard et al., 2003). This method has of course the potential of producing a huge amount of background. Interestingly, experiments with uni-formly ^{15}N-labeling of the overexpressed protein have demonstrated that despite the fact that the entire cell with all its components gets ^{15}N-labeled only a minimal level of background signals is produced (Serber et al., 2001a, b). In contrast, labeling with ^{13}C results in a high level of background signals that makes an unambiguous identification of the signals of the protein of interest impossible with the exception of signals with unique chemical shifts such as high field shifted methyl groups (Wieruszeski et al., 2001; Serber et al., 2004a). This surprising difference in the level of background signals between ^{15}N and ^{13}C is most likely based on the fact that C–H groups are far more abundant in small molecules in the cell than N–H groups. In addition, amide protons that are not protected from solvent exchange through involvement in hydrogen bonds exchange fast with the bulk water, thus making their NMR signals undetectable due to exchange line broadening. Further investigations have, however, revealed that amino acid type selective labeling schemes can produce virtually background free spectra both for ^{15}N and for ^{13}C labeled amino acids (Serber et al., 2001b, 2004a). Unfortunately, not all amino acids can be used in such selective labeling schemes because they are used as metabolic precursors for other amino acid types. Amino acids that are at the end of a metabolic pathway and can thus be used for in-cell NMR experiments are Lys, Arg, or His for ^{15}N labeling and Met and with some restriction Ala for ^{13}C labeling. Other amino acids might also be possible, but however, have not been tested yet. In case other amino acid types should be labeled auxotrophic strains that are incapable of producing certain types of amino acids can be used (Waugh, 1996; McIntosh, 1990). However, the expression yield of such strains is usually reduced, leading to a reduced sensitivity of the NMR spectra.

2.   Applications

Most in-cell NMR applications do not aim at determining the structure of biological macro-molecules directly in the cellular environment, but to use the sensitivity of the chemical shift towards changes in the environment to obtain information about the state of a macromolecule in its natural surrounding or about binding events. One of the main challenges for structure determination of a protein in the cellular environment is that the lifetime of the bacteria or of the protein inside the bacteria is smaller than the measurement time for the average three-dimensional NMR experiment. Recently, however, NMR techniques have been developed that allow spectroscopists to considerably accelerate the measurement of large three-dimensional data sets (Venters et al., 2005; Liu et al., 2005). These reduced dimensionality techniques have already been successfully used in in-cell NMR applications (Reardon and Spicer, 2005). In the future these techniques might enable the complete backbone assignment of a protein in its natural environment. This de novo assignment might be necessary if the chemical shift differences between the in vivo and the in vitro state are too large to transfer the assignment. This will especially be the case for proteins that change their folding state between the crowded cellular environment and a dilute in vitro solution. Such a system has been described by Dedmon et al. who could show that the bacterial protein FlgM which is completely unfolded in vitro is partially folded in the E. coli cytoplasm (Dedmon et al., 2002). In order to prove that the observed differences in the folding states by the protein are due to the high concentration of other (macro-) molecules inside the cell, known as molecular crowding, they added high concentrations of either other proteins (BSA) or small molecules (sugar) to an in vitro sample of FlgM. Comparison of the spectra of the protein obtained with these mixtures with the in-cell spectra showed the same characteristics. In cases like FlgM for which the in vivo situation can be mimicked in vitro, further detailed NMR experiments can of course be carried out in vitro which provide more stable sample conditions.

In-cell NMR experiments are also useful for the investigation of protein-drug interactions. Observing chemical shift differences in HSQC spectra during titration of a protein sample with a potential drug is a widespread application of NMR spectroscopy in the pharmaceutical industry (Fesik, 1993; Shuker et al., 1996; Hajduk et al., 2005; Stockman and Dalvit, 2002). However, the disadvantage of these in vitro screens is that the conditions are quite different from an in vivo situation. A drug, for example, might not be able to cross the cellular membrane or might get metabolized in the cell or bind tighter to other cellular components. In principle, these disadvantages of in vitro screens can – at least partially – be overcome by using in vivo assays for example in-cell NMR experiments for screening. Hubbard et al. have reported an example of such an in-cell NMR screening application (Hubbard et al., 2003). They could show that the drug BRL-16492PA that binds to the bacterial two-component signal transduction protein CheY in vitro also binds to the same protein inside living E. coli bacteria. They based their conclusion on observing virtually identical chemical shift changes in the $[^{15}N,^{1}H]$-HSQC spectrum of CheY upon adding the drug either to a purified in vitro sample or to a slurry of E. coli overexpressing the protein.

Most chemical-shift mapping applications such as protein-drug screens have so far relied on $[^{15}N,^{1}H]$-HSQC experiments due to the high chemical-shift dispersion of the amide protons and nitrogens. As an alternative methyl group based NMR experiments have been employed in the study of protein-drug interactions due to the high sensitivity and involvement of methyl groups in drug binding. The use of methyl groups as indicators for binding is further supported by an investigation which has demonstrated that within a set of 191 crystal structures of protein-ligand complexes 92% of the ligands had a heavy atom within 6Å of a methyl group carbon while only 82% had a heavy atom in the same distance of a backbone nitrogen (Hajduk et al., 2000). In order to test the use of methyl group based $[^{13}C, ^{1}H]$-HSQC experiments in drug screens we have investigated the interaction of calmodulin with the known drug phenoxybenzamine hydrochloride which is assumed to bind to a hydrophobic pocket that is lined with methionines (Serber et al., 2004a). We added the drug to an *E. coli* culture expressing calmodulin half an hour prior to sample preparation. Although no differences in chemical shifts between an in vitro sample and the in-cell sample could be detected, some of the peaks in the in-cell spectrum showed increased line broadening, suggesting that a weak interaction with the drug exists. Further investigations have demonstrated that phenoxybenzamine hydrochloride is mainly associated with the bacterial membrane and that the high local concentration of phenoxybenzamine near the bacterial membrane is most likely responsible for the observed weak interaction.

3. Limitations and Future Directions

The biggest disadvantage of NMR spectroscopy in general is its inherent low sensitivity, making relatively high concentrations of the observed macromolecules necessary. In particular, in-cell NMR investigations that focus on the observation of the behavior of a particular protein in its natural environment are affected by this disadvantage. Most proteins occur in the cellular environment at the low μM to nM level. Interaction studies of an overexpressed protein that reached mM concentrations in the bacterial cyto- or periplasm with its natural interaction partners is, therefore, in most cases based on the current sensitivity of NMR spectrometers impossible. The detection limits for $[^{15}N, {}^{1}H]$-HSQC-based in-cell NMR experiments is approximately 200 μM (Serber et al., 2001b; Hubbard et al., 2003), while the corresponding sensitivity for methyl group $[^{13}C, {}^{1}H]$-HSQC experiments is approximately 70 μM (Serber et al., 2004a). The higher sensitivity of the methyl group NMR experiments is due to the higher number of heteronucleus bound protons, the faster internal rotation of the methyl groups, resulting in slower relaxation and to the fact that amide protons can chemically exchange with the bulk water, which leads to exchange broadening. In-cell NMR experiments can be applied to the investigation of the folding state of proteins, the binding of small molecules that are highly abundant or are added externally or whenever a catalytic relationship between the interaction partner and the protein of interest exists, for example proteins that become phosphorylated of proteolytically processed. Additional applications include the investigation of the overall binding state of the protein in the cellular environment. While proteins like NmerA tumble freely in solution, others like thioredoxin are not visible by in-cell NMR experiments, suggesting that they are strongly interacting with other cellular components that slow the tumbling rate sufficiently to broaden the NMR signals of the protein beyond the detection limit.

The final goal of in-cell NMR experiments is, however, the observation of proteins at or near their physiological concentration. This goal can only be achieved if the sensitivity of NMR spectrometers can be significantly improved. Fortunately, the introduction of cryogenic probes has dramatically increased the sensitivity of NMR instruments over the past years and further improvements are expected. Currently, the biggest challenge for in-cell NMR spectroscopy, however, is its extension to eukaryotic cells. Some preliminary experiments with yeast, insect cells, and in particular with injection of proteins labeled with NMR-active isotopes into Xenopus oocytes (Selenko and Wagner 2004; Serber et al., 2004b) have shown that experiments with eukaryotic cells are in principle possible. However, for sensitive eukaryotic cells such as insect cells or even mammalian cells further improvements of the quality of in-cell NMR experiments and a concomitant reduction in the required overexpression level have to be achieved. For these cell types improvements can be expected from the use of modified NMR tubes that allow for a continuous exchange of oxygenated and nutrient rich media. Such devices have been used in classical in vivo NMR investigations (McGovern, 1994). They will allow researchers to extend the measurement time of the experiments which will further decrease the detection limit. Initial experiments with agarose-trapped bacteria expressing NmerA have been successful (Reckel et al., 2005) and indicate that in-cell NMR experiments with more sensitive eukaryotic cells might be possible in the future.

References

1. Bachert, P., *Progress in Nuclear Magnetic Resonance Spectroscopy,* 1998, 33, 1.
2. Bollard, M.E., Keun, H.C., Beckonert, O., Ebbels, T.M., Antti, H., Nicholls, A.W., Shockcor, J.P., Cantor, G.H., Stevens, G., Lindon, J.C., Holmes, E., and Nicholson, J.K. *Toxicol. Appl. Pharmacol.* 2005a, 204, 135.
3. Bollard, M.E., Stanley, E.G., Lindon, J.C., Nicholson, J.K., and Holmes, E. *NMR Biomed.* 2005b, 18, 143.
4. Cohen, J.S., Lyon, R.C., and Daly, P.F. *Monitoring Intracellular Metabolism by Nuclear Magnetic Resonance,* Vol. 177, Academic Press, San Diego, 1989.
5. Dedmon, M.M., Patel, C.N., Young, G.B., and Pielak, G.J., *Proc. Natl. Acad. Sci. USA,* 2002, 99, 12681.
6. Fesik, S.W., *J. Biomol. NMR,* 1993, 3, 261.
7. Gillies, R.J., *NMR in Physiology and Bionedicine,* Academic Press, San Diego, 1994.
8. Hajduk, P.J., Augeri, D.J., Mack, J., Mendoza, R., Yang, J., Betz, S.F., and Fesik, S.W. *J. Am. Chem. Soc.* 2000, 122, 7898.
9. Hajduk, P.J., Huth, J.R., and Fesik, S.W. *J Med Chem.* 2005, 48, 2518.
10. Hubbard, J.A., MacLachlan, L.K., King, G.W., Jones, J.J., and Fosberry, A.P. *Mol. Microbiol.* 2003, 49, 1191.
11. Kanamori, K. and Ross, B.D. *J. Neurochemistry,* 1997, 68, 1209.
12. Li, C.W., Negendank, W.G., Murphy-Boesch, J., Padavic-Shaller, K., and Brown, T.R. *NMR in Biomedicine,* 1996, 9, 141.
13. Liu, G., Aramini, J., Atreya, H.S., Eletsky, A., Xiao, R., Acton, T., Ma, L., Montelione, G.T., and Szyperski, T. *J. Biomol. NMR,* 2005, 32, 261.
14. McGovern, K.A., in *NMR in Physiology and Biomedicine,* ed. R. J. Gillies, Academic Press, San Diego, 1994, pp. 279.
15. McIntosh, L.P. and Dahlquist, F.W. *Q. Rev. Biophys.* 1990, 23, 1.
16. Reardon, P.N. and Spicer, L.D. *J. Am. Chem. Soc.* 2005, 127, 10848.
17. Reckel, S., Löhr, F., and Dötsch, V. *ChemBioChem.* 2005, 6, 1601.
18. Selenko, P., Wagner, G. in *45th ENC,* Asilomar, Pacific Grove, 2004.
19. Serber, Z. and Dötsch, V., *Biochemistry,* 2001, 40, 14317.
20. Serber, Z., Keatinge-Clay, A.T., Ledwidge, R., Kelly, A.E., Miller, S.M., and Dötsch, V. *J. Am. Chem. Soc.* 2001a, 123, 2446.
21. Serber, Z., Ledwidge, R., Miller, S.M., and Dötsch, V. *J. Am. Chem. Soc.* 2001b, 123, 8895.
22. Serber, Z., Straub, W., Corsini, L., Nomura, A.M., Shimba, N., Craik, C.S., Ortiz. de Montellano, P., and Dötsch, V. *J. Am. Chem. Soc.* 2004a, 126, 7119.
23. Serber, Z., Liu, C., Dötsch, V., and Ferrell, J.E. in *45th ENC,* Asilomar, Pacific Grove, 2004b.
24. Shuker, S.B., Hajduk, P.J., Meadows, R.P., and Fesik, S.W. *Science,* 1996, 274, 1531.
25. Stockman, B.J. and Dalvit, C., *Prog. Nuc. Magn. Reson. Spec.* 2002, 41, 187.
26. Venters, R.A., Coggins, B.E., Kojetin, D., Cavanagh, J., and Zhou, P. *J. Am. Chem. Soc.* 2005, 127, 8785.
27. Waugh, D.S., *J. Biomol. NMR,* 1996, 8, 184.
28. Wieruszeski, J.M., Bohin, A., Bohin, J.P., and Lippens, G. *J. Magn. Reson.* (2001). 151, 118.

PROTEIN-MEMBRANE INTERACTIONS: LESSONS FROM *IN SILICO* STUDIES

ROMAN G. EFREMOV[1*], PAVEL E. VOLYNSKY[1], ANTON A. POLYANSKY[1,2], DMITRY E. NOLDE[1], AND ALEXANDER S. ARSENIEV[1]

[1]*M.M. Shemyakin and Yu.A. Ovchinnikov Institute of Bioorganic Chemistry, Russian Academy of Sciences, Moscow 117997, Russia*
[2]*Department of Bioengineering, Biological Faculty, M.V. Lomonosov Moscow State University, Moscow 119992, Russia*
Corresponding author, E-mail: efremov@nmr.ru

Abstract: Membrane and membrane-active peptides and proteins play a crucial role in numerous cell processes, such as signaling, ion conductance, fusion, and others. Many of them act as highly specific and efficient drugs or drug targets, and, therefore, attract growing interest for biomedical applications. Because of experimental difficulties with characterization of their spatial structure and mode of membrane binding, essential attention is given now to molecular modeling techniques. During the last years an important progress has been achieved in computer simulations of peptides and proteins with various types (implicit and/or explicit) of theoretical models of membranes.

The present work sums up our recent results of molecular dynamics (MD) simulations of binding of several membrane active peptides and proteins to hydrated lipid bilayers and detergent micelles. To check the predictive power of the computational approach, peptides and proteins with diverse folding (α-helical and β-structural), mode of membrane binding, and biological activities were studied. Among them are cardiotoxins (CTXs) from snake venom and fusogenic peptides. The emphasis is made on structural and/or functional information which may be obtained via molecular modeling. In particular, to address the question about the role of membrane composition in protein–lipid interactions, MD simulations were performed in lipid bilayers differing in length, of acyl chains, chemical nature and/or charge of headgroups. It was shown that the results obtained are in a reasonable agreement with experimental data. Possible relationships between the structural/dynamic properties of proteins in different membrane-mimic media and their biological activities are discussed. The approximations and shortcomings of the theoretical models, along with their perspectives in design of new membrane active drugs, are outlined. A general conclusion was reached that *in silico* technologies represent a powerful tool in the field of structural biology of membrane proteins.

1. Introduction

Peptides and proteins which require interaction with membranes to accomplish their functional activities constitute ~30% of the whole genomes [1]. Such molecules are divided into two main classes: some of them permanently reside in the membrane bound state while others are capable of inserting into lipid bilayers from the polar milieu. Various membrane receptors, ion channels and different membrane bound proteins from the first class mediate many

J. D. Puglisi (ed.), Structure and Biophysics – New Technologies for Current Challenges in Biology and Beyond, 19–39.

important cell processes, including molecular and ion transport, intercellular reception and communication via signal transduction. Proteins and peptides from the second class possess a wide range of membrane activities often concerned with alteration of properties of the host membranes. For instance, adsorption of antimicrobial peptides on bacterial membranes leads to cells death due to strong destabilization of their membranes [2]. Another example is the action of viral fusion proteins. Their interactions with eukaryotic cell induce membrane conjugation and entrance of enveloped viruses into cells. It is significant, that functioning of such peripherally bound peptides and proteins is controlled by a large number of factors: their conformation, the mode of membrane binding, the interactions with other proteins or small molecules, pH, the phase condition and lipid composition of membrane [3]. Therefore, the understanding of the structure–function relationship for membrane-active peptides and proteins represents an intriguing challenge in the field of structural biology. Apart from fundamental importance (studies of general principles of protein insertion, folding and stabilization in bilayer), solving the problem is invaluable in optimization of these molecules' behavior for pharmaceutical and biotechnological applications, such as development and targeted delivery of drugs through membranes or action on membrane-bound receptors, design of membrane-active proteins with prescribed properties, gene therapy, and disease control. Unfortunately, studies of biological membrane-protein systems are very difficult because of their complexity. Possible solution of this problem is employment of different membrane mimic systems such as micelles of detergents, lipid vesicles and bilayers. These systems are currently well characterized with the help of experimental methods (see [4] for recent review). However, these techniques often give only overall picture of a model peptide-membrane system, while molecular details are missing. Development of molecular modeling approaches would be indispensable to avoid these problems. Such methods are widely used in studies of protein-membrane interactions (reviewed in [5–8]). Molecular dynamics (MD) of peptides and proteins in explicit membrane environments (lipid bilayers and micelles) is one of the most powerful among them. This method permits atomic-scale studies of protein interactions with lipids and detergents, changes in protein structure upon binding, and "membrane response" induced by protein insertion.

The present work sums up our recent results on MD simulations of binding of several membrane active peptides and proteins to hydrated lipid bilayers and detergent micelles. To check the predictive power of the computational approach, peptides and proteins with diverse folding (α-helical fusogenic peptides and β-structural cardiotoxins (CTXs) from snake venom), mode of membrane binding, and biological activities were studied. The emphasis is made on structural and/or functional information which may be obtained via molecular modeling. In particular, to address the question about the role of membrane composition in protein–lipid interactions, MD simulations were performed in lipid bilayers differing in length, of acyl chains, chemical nature and/or charge of headgroups.

CTXs from snake venom belong to the family of "three-finger" (or "three-loop") small β-sheet proteins rich in disulfide bonds which are capable of binding to cell membranes and damage a wide variety of cells presumably by perturbing the structure of lipid bilayers (e.g. [9]). Experimental studies reveal that spatial structures of CTXs do not change significantly upon interactions with membrane-mimic medium. On their other hand, minor conformational differences may lead to dramatic alterations of their activity [10]. To explore the role of structural heterogeneity in membrane binding, we performed MD simulations of the CTX I (A1) from the venom of cobra *Naja oxiana*. This was done in water, DPC and SDS micelles. The high-resolution three-dimensional (3D) structure of A1 has been recently solved in aqueous solution and DPC micelles by NMR spectroscopy [11, 12]. Therefore, the computational results can be directly compared with the experimental data.

Attachment of enveloped viruses to the target cell is mediated by anchoring of fusion proteins on the membrane surface via their conservative fragments, "fusion peptides" (FPs), consisting of about 20 amino acid residues (see [3] for recent reviews). As a result, FPs insert into the host membrane, providing close contact between lipid bilayers of the virus and the cell. FP of Influenza A virus hemagglutinin (HA) is one of the most promising objects for studies of molecular events that occur in cell membranes upon binding and insertion of FPs. Also, fusion and leakage activities have been reported for a number of HA-homologs [14]. The spatial structures of HA and its active water-soluble analog, peptide E5, have been solved by NMR spectroscopy in detergent micelles [14, 15]. Both peptides peripherally bind to the membrane surface and insert into the hydrophobic medium with their amphiphilic N-terminal α-helices, while the C-terminal parts are rather flexible. One important challenge in studies of the mechanisms of FPs' action is the peptide-induced "membrane response". This, in turn, depends on the lipid composition of the membrane. In this study such effects were explored via MD simulations of the peptide E5 in two full-atom hydrated lipid bilayers composed of DMPC and DPPC – two lipids differing in the length of their acyl chains.

2.   Computational Details

Information about the membrane active agents and parameters of simulated systems are presented in Tables 1 and 2.

TABLE 1. Simulated membrane active agents

Protein/peptide	Sequence	Net charge	Initial conformation	Source
Cardiotoxin A1 from *Naja oxiana*	LKCNKLVPIAYKTCPEGKNL CYKMFMMSDLTIPVRKGCID VCPKNSLLVKYVCCNYDRCN	+6	β-structural	NMR in water/DPC mixture (1ZAD) [12]
Fusion peptide E5	GLFEAIAEFIEGGWEGLIEG	–5	α-helical	NMR in water/DPC mixture [15]

2.1.   SYSTEMS PREPARATION

2.1.1. Bilayers and micelles

Initial conformations of detergent and lipid molecules were generated in the molecular editor. Detergent molecules were collected into the micelle model with predefined aggregation numbers and uniform distribution of polar heads on the micelle surface. The aggregation numbers were taken from the experimental data [16, 17]. Starting configurations of lipid bilayers (128 lipids + SPC waters) were constructed in such a way that their parameters (area per lipid molecule, bilayer thickness, water/lipid ratio) were close to the experimental ones. Total charges of the systems were neutralized by adding of necessary numbers of counterions.

The resulting systems were subjected to energy relaxation via 5×10^4 steps of steepest descent minimization followed by heating from 5 K to the target temperature (Table 2) during 50-ps MD run. Then the long-term collection MD runs were carried out. The final configurations of bilayers and micelles were further used in MD simulations of their complexes with proteins and peptides. Other simulation details for pure lipid bilayers are given elsewhere [18, 19].

2.1.2. Protein and peptides in membrane environment

Because of limitations of the simulations timescale (<30 ns), realistic choice of the starting configuration is very important. This was done based on the results of Monte-Carlo (MC) conformational search with implicit membrane model. Previously, this approach has been successfully applied to study membrane binding of a number of α-helical and β-structural proteins and peptides [8]. In this work the initial orientation of A1 with respect to micelles was that obtained via MC simulations in implicit membrane starting from the structure solved by NMR spectroscopy in DPC/water mixture [11]. Details of MC simulations were described elsewhere [20].

The starting conformation of the peptide E5 was that obtained by NMR spectroscopy in DPC micelles [15]. Its coordinates were kindly provided by Dr. P.V. Dubovskii. In simulations of E5 with lipid bilayers (DMPC and DPPC) the starting systems were constructed as follows. First, the water molecules were removed from the equilibrated box, and the peptide was placed at a distance ~10 Å from the bilayer interface – with its hydrophobic residues oriented toward the membrane and the hydrophilic ones facing the water phase. Next, the systems were solvated, and five counterions were added. The system was then subjected to energy minimization and subsequent heating during 50 ps. Finally, 2-ns MD simulations were carried out with an external acceleration of 1.0 Å/ps² applied to the peptide atoms along the normal to the membrane

TABLE 2. Parameters of the systems under study and MD simulation protocols

System[a]	T (K)	Box ($Å^3$)	Electrostatics treatment[b]	Time (ns)
Pure micelles				
$DPC_{60}/water_{17385}$	323	$83.0 \times 85.3 \times 80.0$	Cutoff	10
$SDS_{60}/water_{7811}/Na^+_{60}$	323	$63.6 \times 63.3 \times 64.9$	PME	10
Pure bilayers				
$DMPC_{128}/water_{3303}$	325	$61.8 \times 61.8 \times 62.7$	Cutoff	30
$DPPC_{128}/water_{3955}$	325	$61.0 \times 61.9 \times 72.5$	Cutoff	20
Proteins and peptides in water				
$A1/waters_{13738}/Cl^-_6$	323	$79.6 \times 67.4 \times 80.6$	Cutoff	10
$E5/water_{7043}/Na^+_5$	305	$60.0 \times 60.0 \times 60.0$	Cutoff	15
Proteins and peptides in membrane environment				
$A1/DPC_{60}/water_{24137}/Cl^-_6$	323	$80.0 \times 100.5 \times 97.5$	Cutoff	20
$A1/SDS_{60}/water_{15935}/Na^+_{54}$	323	$80.8 \times 80.8 \times 80.8$	PME	20
$E5/DMPC_{128}/water_{7219}/Na^+_5$	325	$61.8 \times 61.8 \times 95.0$	Cutoff	20
$E5/DPPC_{128}/water_{5188}/Na^+_5$	325	$61.3 \times 61.8 \times 83.1$	Cutoff	20

[a] Subscripts indicate the number of molecules or ions in a system.
[b] Cutoff – spherical cutoff function for nonbond interactions, PME – electrostatic interactions were treated using the Paticle-mesh Ewald algorithm.

plane. At this stage, a number of NMR-derived geometrical restraints were employed to preserve initial conformation of the peptide. The following parameters were analyzed for all peptide/bilayer systems: (i) The depth of peptide insertion into the membrane, (ii) the energies of peptide-membrane interactions, (iii) the peptide's secondary structure, (iv) RMSD of the peptide conformations from the NMR structure, (v) H-bonding. The resulting states (revealing strong peptide-membrane interaction and small deviation from the initial peptide conformation (root-mean-square deviations, RMSD <1 Å)) were taken as the starting structures for subsequent unrestrained MD runs.

Furthermore, to assess the role of media effects approximating membrane environment, MD-simulations of all the polypeptides were carried out in explicit water.

2.2.   SIMULATION PROTOCOLS

All simulations were performed using the GROMACS 3.2.1 [21] package and the force field specially adopted for lipids [22]. MD Simulations were carried out with a time step of 2 fs, with imposed 3D periodic boundary conditions, in the NPT ensemble with isotropic pressure of 1 bar. Van der Waals interactions were truncated using the twin range 12/20 Å spherical cutoff function. Electrostatic interactions were treated in two different ways: by use the same cutoff scheme and the PME algorithm (these questions were discussed in [18]). The simulation temperatures (Table 2) were selected to ensure liquid crystalline state of lipid bilayers. Other computational details were described elsewhere [18, 19].

2.3.   ANALYSIS OF MD-TRAJECTORIES

MD trajectories were analyzed using original software developed by the authors and utilities supplied with the GROMACS package. Energies of electrostatic and van der Waals interactions between various components of the systems were calculated using the g-nbi program, specially written for this. The following macroscopic parameters of the simulated systems were analyzed: the area per lipid molecule (A_L), the mean order parameter of acyl chains of lipids (S_{CD}), the distance between phosphorus atoms of different bilayer leaflets (D_{PP}), radius of gyration (R_G) and asymmetry parameter (α) for micelles, the root-mean-square fluctuations (RMSF) of coordinates of heavy or C_α atoms of peptides and proteins for each residue near their average equilibrium positions, the position of peptide's residues relative to the membrane, the energies of interactions of peptides and proteins residues with the membrane or micelle, the density distributions of phosphorus atoms and water along the normal to the membrane plane, the distribution of S_{CD} values over the acyl chains carbon atoms. All these parameters were averaged over the equilibrium parts of corresponding MD trajectories (last 5 ns). Hydrophobic properties of the peptides and proteins were calculated and visualized using the molecular hydrophobicity potential (MHP) approach as described elsewhere [23].

3.   Results and Discussion

3.1.   CARDIOTOXIN A1 IN DETERGENT MICELLES

All known CTXs share the same spatial fold – a five-strand β-sheet stabilized by four disulfide bridges (Figure 1A). Because of such a tight structure, rich of intramolecular H-bonds, they preserve this fold in a wide variety of different environments, such as water, nonpolar solvents, and water-detergent mixtures. CTXs cause cell death by destruction of its membrane. The first stage of their action is the binding to the membrane surface. So, understanding of the factors important for stability of toxin-membrane complexes may shed light on CTXs' activity. Recently, interaction of CTX A2 from *N. oxiana* with the implicit membrane was explored via MC simulations [20]. It was shown that in the membrane-bound state the toxin's surface reveals an extensive hydrophobic pattern spatially edged with a belt of positively charged residues. Being useful in delineation of an overall mode of CTX's binding, simulations in implicit membrane do not provide atomic-scale details of CTX-membrane interactions. This may be done using modeling of CTXs in full-atom membrane-mimicking environments. Here we present MD analysis of CTX A1 from *N. oxiana* in three different environments: in water solution (MD$_W$) and in the presence of DPC (MD$_D$) and SDS (MD$_S$) micelles. As the experimental information about structures of A1 in water and water-DPC mixtures is available, it is possible to check the validity of simulation results. Furthermore, the detergents chosen are rather different, thus permitting assessment of the influence of various properties of micellar environment (charge, packing density of detergents, hydrophobic properties of micelles, Figure 1B–C) on the binding.

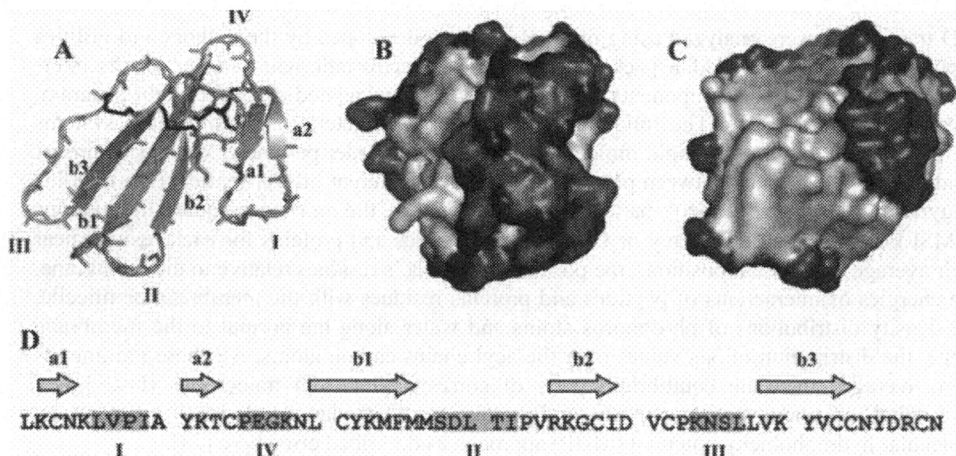

Figure 1. Starting conformations of the toxin A1 (**A**), DPC (**B**) and SDS (**C**) micelles. Side chains of residues are removed for clarity. Loop regions are marked with roman numbers. Hydrophobic and hydrophilic surface areas are colored with light- and dark-gray, respectively. **D.** Amino acid sequence of A1. β-Strands are indicated with *arrows*, flexible regions I–IV are hatched.

3.1.1. Stability of the toxin's structure

Analysis of MD data shows that independently of environment the toxin structure is highly conserved. Main conformational changes occur during the first 5 ns of MD. At this stage RMSD from the starting structure attains to its maximum value (3.3 Å in MD_W, MD_D, and 2.5 Å in MD_S) and then remains unchanged. Secondary structure of A1 is well preserved: both β-sheets exist all the simulation time, although some minor changes in individual β-strands occur. Namely, shortening of the β-strand b2 is detected in all MD trajectories. Also, in the presence of micelles the structure of the loop III is stabilized by formation of either a β-bridge (MD_S) or a short β-strand (MD_D). The system composition greatly affects on the stability of the secondary structure elements during the simulation: the length of β-strand b2 oscillates near its mean value with a period ~2 ns in MD_W, while in MD_D and MD_S the secondary structure is stable in equilibrium.

Relative stability and flexibility of the protein residues were estimated based on the values of their RMSFs from average positions during the last 5 ns of MD (Figure 2). It is seen that in all MD the most flexible regions are the loops I–III and residues 15–19 (IV). These data are in a good agreement with the NMR studies, which also revealed large conformational variability of these regions. Presence of micelle significantly decreases conformational lability of the toxin: RMSF values for the residues in loops I–III are less than 1 Å in MD_D and MD_S, while in water they exceed 2.5 Å. Region IV is also more stable in micelle environment: RMSFs for these residues are equal to 1.6 Å in MD_S, and 1 Å in MD_D (in water RMSF ~2.5 Å). So, the presence of micelles greatly stabilizes the toxin's structure.

To summarize, we can conclude that during MD the structure of A1 does not undergo strong deviations from the starting one. The estimated dynamic characteristics of the toxin residues correspond to the experimental data – the residues with high structural variability in NMR–derived models possess the highest flexibility in MD calculations. The main influence of micelle environment consists in stabilization of A1 loops regions. Possible reason for this is the interaction with the micelles, which will be analyzes in the next section.

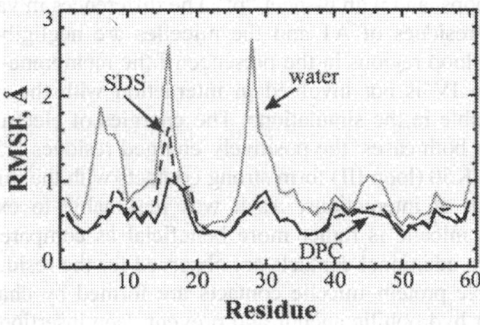

Figure 2. Conformational flexibility of the toxin's residues (expressed in terms of RMSF) in different environments: in water (*gray*), DPC (*solid line*), and SDS (*dashed line*) micelles. RMSF values are calculated on the equilibrium part of MD trajectories.

3.1.2. Protein–environment interactions

One of the global characteristics of protein interactions with environment is the nonbond energy. In MD_W energies of protein interactions with water and counterions remain invariable during all the simulation time. This indicates that equilibration of the system is rather fast (at least according to such criterion). Another picture is observed in the presence of micelles. In these cases the energies of van der Waals interactions with water slightly increase (by ~100 kJ/mol) on the first stage of MD (10 ns in MD_D and 4 ns in MD_S). Such energy loss is compensated by significant drop of the interaction energy with micelles (from –200/–400 to –600/–800 kJ/mol). The same tendency is found for the electrostatic term – the energy of interactions with micelle is decreased due to the loss of contacts with water. The gain in electrostatic interactions is more pronounced in charged SDS micelle (~7000 kJ/mol) even despite unfavorable contacts between detergent molecules and counterions (~4000 kJ/mol). So, the total energy gain in SDS micelle is ~3000 kJ/mol. In a case of DPC micelle this value is equal to ~2000 kJ/mol. We should note that the total electrostatic energy of the toxin's interactions with environment in both cases stays unchanged during all the simulation time. So, we can conclude that in both cases protein–micelle interactions reach equilibrium after ~10 ns of MD. At the first stage, redistribution of protein contacts with water and micelles is observed, thus leading to formation of a tightly-bound toxin-micelle complex. It is stabilized by both electrostatic and van der Waals interactions. Appearance of favorable electrostatic contacts with micelles compensates the loss of interactions with water. Van der Waals interactions with DPC micelle are more favorable then with SDS one. The same tendencies are observed in redistribution of hydrogen bonds with the environment – the number of H-bonds with micelles significantly increases simultaneously with the loss of H-bonding with water. The number of intraprotein H-bonds remains unchanged in both cases.

To delineate the binding motif in the toxin molecule, we performed analysis of protein–micelle interaction energies for individual residues. As seen in Figure 3, regardless of the micelle the mode of protein binding is the same. The toxin contacts the micelle with the hydrophobic tips of its loops, and with β-sheet "b". The differences in van der Waals energies of interactions between residues of A1 and the micelles are negligible. This explains the increased stability of the loop regions in the presence of the membrane-mimics. On the other hand, the flexible region IV is not involved in interaction with the micelles. Indeed, this protein part is quite flexible in the simulations. The energies of electrostatic interactions of residues also correlate in both cases: the positively charged residues K5, K12 (loop I), K24, K35, R36 (loop II), K44, K50 (loop III) form strong contacts with the micelles. Here the main difference is the strength of interactions. As it was reasonable to expect, binding to the negatively charged SDS micelle is rather more beneficial as compared to the zwitterionic DPC micelle. In MD_S two additional contacts are observed for the residues K18 and R58. All the energetically favorable protein–micelle contacts are formed by charged residues. These residues act as a peculiar hydrophilic anchor and prevent deep insertion of the toxin into the hydrophobic interior. So, the binding mode which had been predicted in simulations with the implicit membrane [20] is well reproduced in full-atom DPC and SDS micelles. This justifies combined employment of both types of membrane models and demonstrates the importance of such a structural feature for binding of CTXs to membrane mimics.

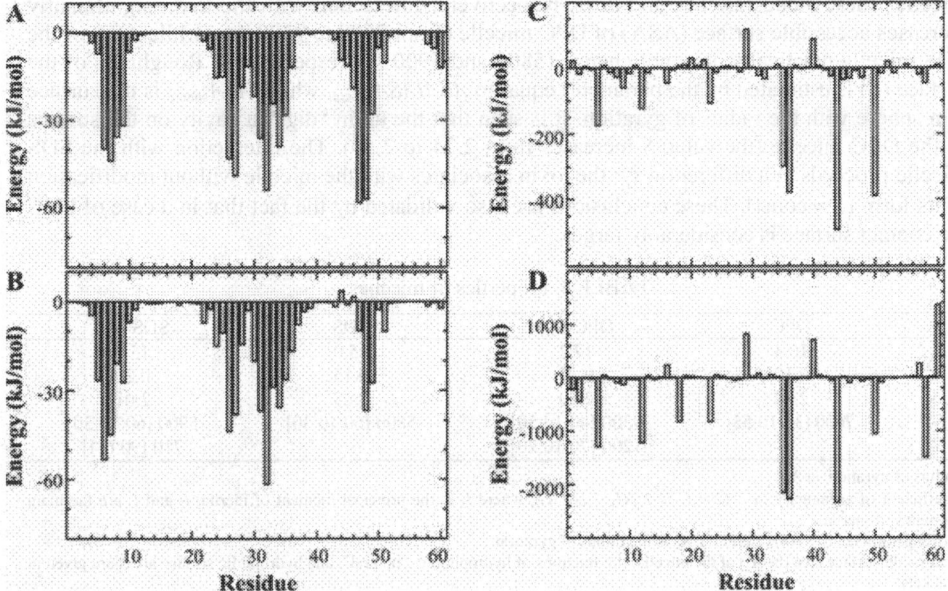

Figure 3. Interaction of cardiotoxin A1 with micelles. Van der Waals (**A, B**) and electrostatic (**C, D**) energies of interaction of the toxin's residues with DPC (**A, C**) and SDS (**B, D**) micelles. The data are averaged over equilibrium parts of MD trajectories.

3.1.3. Micelles' properties

Binding of toxin to micelles induces changes of structural and dynamic parameters of the latter ones (geometry, packing, order parameters, etc.). Corresponding values calculated on the equilibrium part of MD trajectories are shown in Table 3. It is seen that the studied micelles are quite different. SDS micelle is much smaller, has an asymmetric form, and is relatively smooth. Despite the charge of its polar groups, its surface is mainly hydrophobic. These properties are explained by the character of interactions between detergent molecules in the micelle. The polar heads of SDS are rather small, so their total surface is not enough to cover the micelle. Polar groups of SDS form on the micelle surface clusters connected by a network of salt bridges through the Na^+ ions. The acyl chains of SDS adopt mainly extended conformation and reveal strong van der Waals contacts with each other. DPC micelle is bigger, more spherical, and its surface is rough. Its surface is much more polar than that in the SDS one. The polar head of DPC is greater and consists of two groups of different charge. These groups are included into a network of electrostatic contacts on the micelle surface, but these interactions are weaker than those in the SDS micelle. The acyl chains of DPC are rather disordered. Binding of A1 to the micelles strongly depends on the properties of the latter ones. Since the SDS micelle is more rigid, its deformation upon the toxin insertion is weaker. Indeed, its size and asymmetry increase only by 2% and 30%, while in a case of DPC micelles these values are 5% and 80%, respectively. Interaction with the toxin significantly increases accessible surface (ASA) of DPC micelle (from 7600 to 9300 Å^2), while ASA of the SDS micelle practically does not change (5800 and 5900 Å^2, respectively). Roughness of the surface (S) is estimated by the parameter equal to ASA/ASA_{sphere}, where ASA_{sphere} is the surface of a sphere with the radius of gyration. It is seen that the toxin "digs" a cavity on the surface of the DPC micelle (the value S increases from 2.24 to 2.50). The interaction with the SDS micelle proceeds in a different way: the toxin associates with the micelle without modification of its form ($S \approx$ const). These conclusions are also validated by the fact that in a case of DPC the contact surface is considerably larger.

TABLE 3. Properties of micelles

	DPC	DPC+A1	SDS	SDS+A1
$R_G{}^a$	16.4	17.2	15.0	15.3
α^b	0.10	0.18	0.15	0.20
S^c	2.24	2.50	2.05	2.01
$S_T{}^d$	7600 [35:10:55]	9300 [40:10:50]	5800 [60:10:30]	5900 [60:10:30]
$S_C{}^e$		1200 [70:10:20]		710 [90:6:4]

^{a}Radius of gyration, in Å.
^{b}Coefficient of asymmetry, $\alpha = (2 \times I_1-I_2-I_3)/(I_1 + I_2 + I_3)$, where I_1 is the principal moment of inertia, I_2 and I_3 are the main moments of inertia.
^{c}Ruffle parameter, $S = ASA/(4\pi RG^2)$, R_G is the radius of gyration.
^{d}Accessible surface area (in Å^2) of the micelle, the fractions of hydrophobic, "neutral", and hydrophilic surface areas are given in brackets.
^{e}S_C is the contact area.

To gain additional insight into the mechanisms of the toxin binding to micelles, we analyzed the spatial hydrophobic properties of the contact surface. For this purpose, the MHP (e.g. [23]) induced by the detergent atoms was calculated in the points of the micelle surface. As seen in Table 3, in both micelles the contact surface is rather more hydrophobic then the other surface zones. The hydrophilic part of the contact surface represents only 20% in DPC micelle and 4% in SDS micelle. Inspection of the toxin's residues contacting with the micelle shows that they are mainly hydrophobic, except the residues in the "polar belt". The hydrophilic regions of the micelle surface fall in contact with charged protein residues. Therefore, the interfacial region between the toxin molecule and the micelle reveals prominent complementarity of hydrophobic and hydrophilic properties of the interacting molecules. Interestingly, similar "3D hydrophobic match" effect is also observed for α-helical peptides in the membrane-bound state (see below).

3.2. FUSION PEPTIDE E5

Binding of FPs to lipid bilayers, as well as their functional activity in the membrane-embedded state essentially depend on the physicochemical characteristics of the membrane. Previously, MD simulations of FPs in full-atom hydrated lipid bilayers have been carried out to investigate the conformation of HA in dimyristoyl phosphatidylcholine (DMPC) bilayer [24], showing that the peptide binds on the water-membrane interface and adopts a kinked conformation. This agrees well with the NMR data [14]. Moreover, the influence of HA on the bilayer structure was detected. In particular, it was found that the average hydrophobic thickness of the lipid phase near the N-terminus of HA is reduced as compared to that in pure bilayer. The peptide HA and a set of its fusogenic and nonfusogenic analogs were also investigated by MD simulations in palmitoyloleyl phosphatidylcholine (POPC) bilayer [25], demonstrating that the N-terminal α-helix of the active analogs of FP is buried into the membrane with its N-terminus and forms an angle ~30° with the bilayer surface. Decreasing of bilayer thickness and destabilization of acyl chains of lipids caused by FP were also observed. The aforementioned MD-results provided for the first time a wealth of interesting microscopic details related to behavior of the peptide HA in two different membranes. On the other hand, several important aspects of FP-membrane interactions were not addressed in these studies. Among them one can outline the following: (i) What are the microscopic details of peptide and water interactions with lipid headgroups in membranes of different composition? (ii) Are there any correlations between 3D hydrophobic properties of a peptide and a bilayer (3D hydrophobic match effects)? With this aim in view, in the present work we carried out long-term MD simulations of the fusion-active peptide E5 in two full-atom hydrated lipid bilayers composed of DMPC and dipalmitoyl phosphatidylcholine (DPPC). Special attention was given to the atomic-level details of the mutual influence of the peptide and the membrane on their behavior. These lipids were chosen for the following reasons: (i) They have similar headgroups and possess minor differences in length of their acyl chains (two CH_2 groups); (ii) Despite that, the corresponding bilayers demonstrate rather different physicochemical properties [26]. Thus, being similar in a packing density (close values of surface area per lipid), the DMPC bilayer is thinner and has larger coefficient of lateral diffusion (i.e., it is more "fluid") as compared to DPPC. Therefore, the specific features of peptide-membrane interactions (if any) should mainly be caused by structural and dynamic characteristics of these lipid-water systems, rather than chemical nature of their particular groups. The main results of MD simulations of E5 in DMPC and DPPC bilayers are discussed below.

3.2.1. Lipid bilayer significantly promotes secondary structure of the fusion peptide

Analysis of MD data obtained for the peptide E5 in DPPC and DMPC bilayers clearly shows that the presence of the water–lipid interface considerably stabilizes α-helical conformation on the N-terminus of the peptide (Figure 4C–D). In both bilayers the residues 2–10(11) form a stable α-helix during all the simulation time. These results agree well with the data derived from the analysis of a set of twenty 3D structures of the peptide obtained by NMR spectroscopy in DPC micelles (Figure 4A). The only exception is that the helical segment in NMR models is shifted by one residue toward the C-terminus. Other important characteristics is the flexibility of different peptide parts on the water-membrane surface. Analysis of C_α RMSF values shows that in the presence of bilayers, the peptide has quite a rigid structure with RMSFs ~0.5 Å (Figure 5A). The exceptions are provided by the C-terminal residue G20 in both membranes and by the residue G16 in DPPC. Therefore, the peptide backbone in both bilayers is strongly conformationally restrained independently of its secondary structure (Figure 4C–D): α-helical, β-turn, and coil regions of the peptide demonstrate equally low values of RMSF. This indicates that stabilization of a peptide on the membrane surface may be reached not only due to formation of intramolecular H-bonds (as in α-helix and β-turn), but can also be induced by strong contacts of completely or partly disordered peptide with the water–lipid interface. At the same time, unlike the DPPC membrane, the charged residues

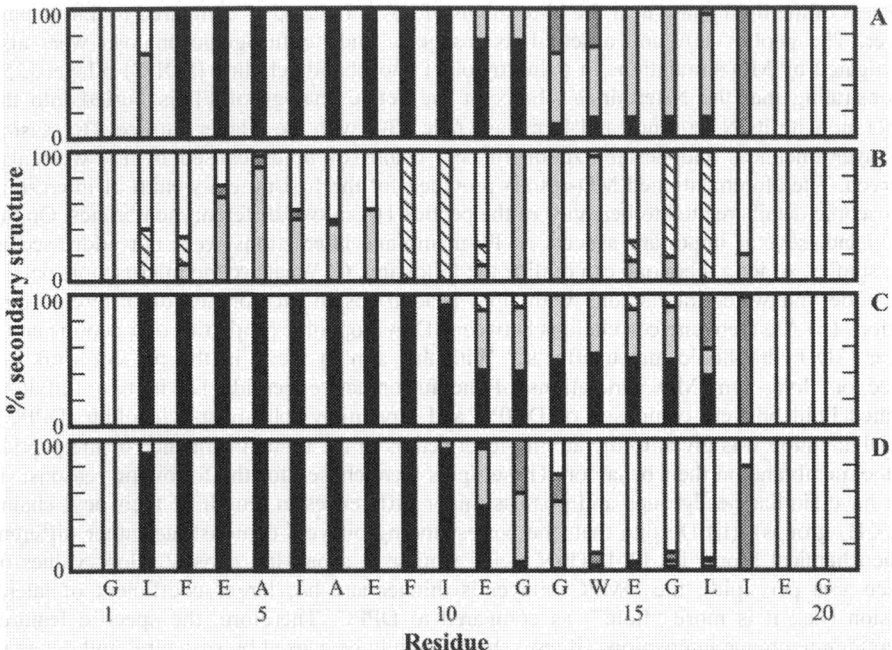

Figure 4. Secondary structure of the peptide E5. Relative occurrence of the peptide's residues in different elements of secondary structure: black, α-helix; light-gray, β-turn; dark-gray, bend; white, random coil; hatched, β-sheet. **A**. NMR-derived models of the peptide E5 in DPC micelles. **B–D**. The results obtained during the last 5 ns of MD simulation of E5 in water (**B**) and with the presence of DMPC (**C**) and DPPC (**D**) bilayers, respectively.

E8, E11, E15 in DMPC reveal high RMSFs for heavy atoms (Figure 5A). Therefore, the lability of these glutamic acids in DMPC is related to behavior of their side chains. (Obviously, the lability of G16 in DPPC and G20 in both systems may be caused only by movements of their backbone atoms.) Possible reasons for such an unexpectedly high difference in mobility of side chains of residues E8, E11, E15 in DMPC will be discussed later.

To assess the influence of membrane on structure and dynamics of the fusion peptide, MD simulations were also carried out in aqueous solution. The results show that behavior of the peptide in water and in the presence of membrane differs dramatically. Thus, after 8-ns MD in water the peptide completely loses its initial α-helical structure and adopts a tightly packed conformation (the values of accessible surface area are 1200 Å^2 in water and 1300–1350 Å^2 on the interface), where only few residues are involved in formation of the secondary structure elements (Figure 4A).

To summarize, we can conclude that, unlike the aqueous solution, the water-membrane interface significantly promotes structuring of the anchored amphiphilic peptide. As discussed later, this facilitates its insertion into lipid bilayer. Such effects are well-known from experiments [27]. In addition, in the membrane-bound form the peptide is highly conformationally constrained as compared to its water-soluble state. The lipid composition of the membrane does affect the structural and dynamic characteristics of the fusion peptide – overall, it is more structured in the more "fluid" (DMPC) membrane. Also, being adsorbed on the DMPC bilayer, the peptide reveals high flexibility of side chains of several negatively charged residues.

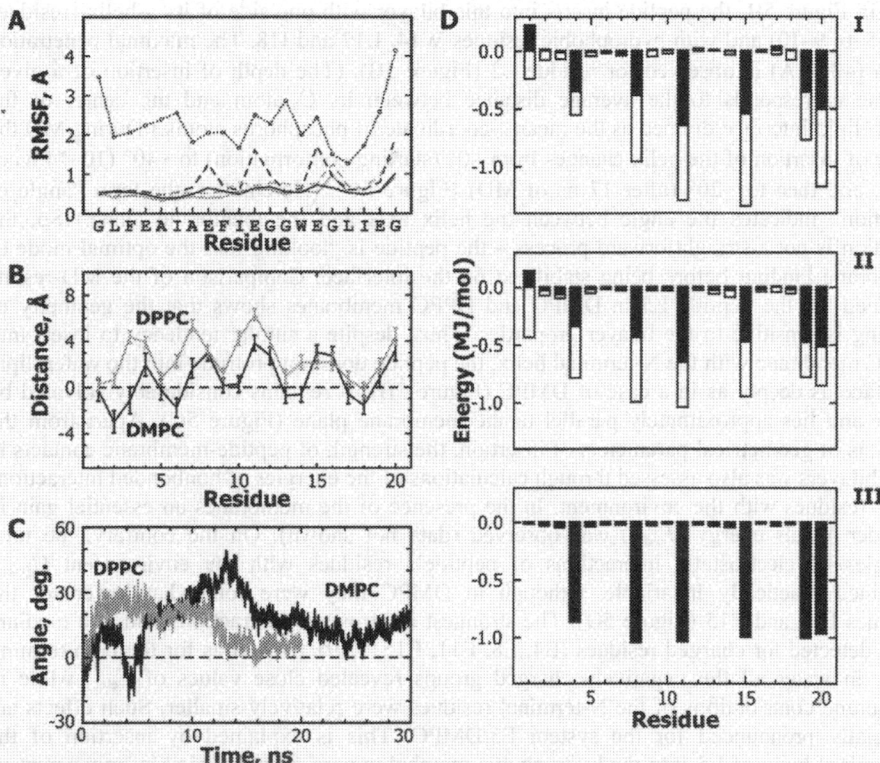

Figure 5. Interaction of the peptide E5 with water (*dotted line*), DMPC (*black*) and DPPC (*gray*) bilayers. **A**. Root-mean-square fluctuations (RMSF) of the peptide heavy (*dashed or dotted line*) and backbone (*solid line*) atoms for each residue. **B**. Depth of insertion of the peptide's residues into the membranes. The distance between C_α-atoms of E5 and the center of water–lipid interface (calculated as a mean z-coordinate of phosphorus atoms). **C**. Time evolution of the angle between the helix axis (calculated between residues 2 and 11) and the bilayer plane. **D**. Energies of electrostatic interactions of the peptide E5 with the environment. MD data averaged over the last 5 ns of simulations. (I) DMPC/E5; (II) DPPC/E5; (II) E5/water. The contributions of water and lipid molecules are indicated with *black* and *white* bars, respectively.

3.2.2. The modes of peptide insertion into lipid bilayers

MD simulations show that the peptide forms stable complexes with both lipid bilayers. It penetrates into the membranes via the N-terminal α-helix, while the C-terminal fragment stays adsorbed on the bilayer surface (Figure 6). Detailed characteristics of the binding are presented in Figure 5. First, let's consider the behavior of E5 in the DMPC membrane. As seen in Figure 5B, the peptide inserts into this bilayer with one side of its α-helix (residues 1–3, 5–6, 9–10) and with hydrophobic residues W14, L17 and I18. The maximal penetration depth ($\sim$3–4 Å) is observed for residue L2 (Figure 5B). (The depth of insertion of a given residue corresponds to the average distance between its C_{α}-atom and the center of the water–lipid interface defined as the mean z-coordinate of phosphorus atoms.) During MD the angle of insertion of the helix changes from $\sim$0° (starting conformation) to $\sim$40° (10–14 ns of MD), and then to $\sim$20° (after 17 ns of MD) (Figure 5C). (Hereinafter, the term "angle of insertion" indicates the angle between the helix axis and the membrane plane.) So, the insertion is not a straightforward process – the peptide is "looking for" the optimal mode of membrane binding before being stabilized on the interface. Comparison of the MD-results obtained for the peptide E5 in DMPC and DPPC membranes shows that the geometry of binding is sensitive to the bilayer properties. Thus, despite a similar tendency to insert into DPPC membrane with the N-terminal helix, the peptide does not protrude into the water–lipid interface as deeply as in a case of DMPC (Figure 5B). It remains considerably solvated by water and lies approximately parallel to the membrane plane (Figure 5C). Apart from the analysis of geometrical parameters of insertion, the strength of peptide-membrane contacts in both bilayers was also assessed through calculations of the energies of nonbonded interactions of its residues with the environment. In the presence of the membranes an essential gain in van der Waals energy (E_{vdW}) was observed (data not shown). On the contrary, the total energies of electrostatic interactions of peptide's residues with the environment ($E_{elec.}$) remained practically invariable, although in DMPC they were somewhat lower for the residues E11 and E15 (Figure 5D). The strongest electrostatic interactions with the medium were detected for charged residues (E4, E8, E11, E15, E19), as well as for the both termini. Also, in water all the negatively charged groups revealed close values of E_{elec}, while in membrane contributions of the N-terminal residues were relatively smaller. Such effects are especially pronounced for the system E5/DMPC. This is explained by insertion of the N-terminal helix of E5 into the hydrophobic membrane core, accompanied by movement of the peptide charged groups away from the polar phase of the system.

The bilayer composition seriously affects the values of E_{elec} for individual residues. The first difference appears in interaction of the N-terminal amino group with the environment: being almost negligible in DMPC (the values of E_{elec} describing contacts with lipids and water have opposite signs), in DPPC the NH_3^+ group is accommodated in a favorable lipid environment (negative values of E_{elec}). For E5 in DPPC bilayer the contributions of lipids and water molecules into $E_{elec.}$ are quite similar. By contrast, in a case of E5/DMPC system the role of electrostatic interactions between the peptide and the polar headgroups of lipids (mainly cholines) becomes more important, especially for residues E11, E15, E19, and for the C-terminal G20. This is caused by two reasons: (i) deeper insertion of E5 into the DMPC bilayer as compared with the DPPC one (Figure 5B); (ii) protruding of some lipid molecules around the binding site into the water phase (Figure 6B). These lipids have favorable contacts with side chains of glutamic acids of the peptide. On the contrary, upon the peptide insertion the DPPC bilayer retains well its integrity. In this case side chains of Glu residues lie above

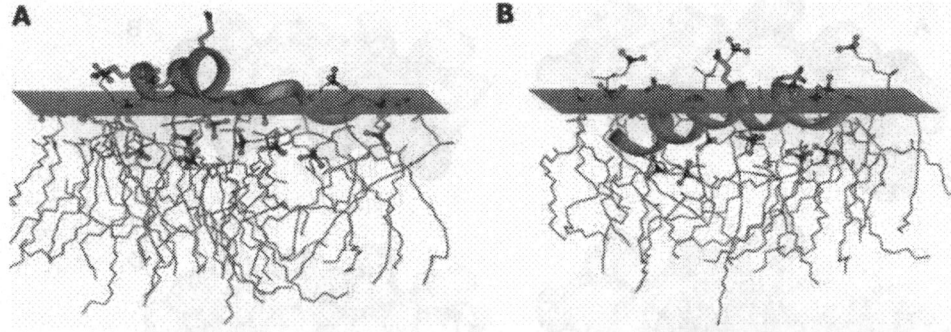

Figure 6. Mode of binding of the peptide E5 to DPPC (**A**) and DMPC (**B**) bilayers. The peptide is given in a ribbon presentation, side chains of glutamic acids are drawn with sticks. The water–lipid interface (calculated as a mean z-coordinate of phosphorus atoms) is schematically shown with gray-hatched plane. Water molecules are removed for clarity. Neighboring lipid molecules are indicated, their choline groups are given in ball-and-stick presentation.

the membrane surface and are rather less accessible to lipids (Figure 6A). We assume that the high flexibility (RMSF values, Figure 5A) of side chains of E8, E11, E15 in DMPC is explained by their strong contacts with individual lipid molecules "plucked" from the bilayer into the water phase, while in the DPPC membrane the lipids are much more structured keeping their positions inside the bilayer.

3.2.3. Peptide-membrane binding: complementarity of the 3D hydrophobic properties

It is well established that binding of peptides on the water-membrane surface, as well as their insertion into nonpolar core of lipid bilayers, is often mediated by hydrophobic interactions [28]. Such systems reveal the prominent dominance of hydrophobic/hydrophobic and hydrophilic/hydrophilic contacts, while the hydrophobic/hydrophilic ones are too energetically costly to be frequently observed. For instance, "hydrophobic match/mismatch" effect, which reflects the correspondence between the length of hydrophobic membrane-spanning regions of a peptide and the thickness of a nonpolar membrane layer – is considered to be one of the most important factors driving peptide-membrane association (see [29] for review). In order to estimate the role of similar effects in binding of the fusion peptide E5 to model membranes, we carried out detailed analysis of spatial hydrophobic properties of the peptide and the lipid bilayer in the vicinity of their closest contacts. The main objective was to check whether the hydrophobic and hydrophilic patterns on the surfaces of the peptide and the bilayer complement each other or not. The term "complement" means that in the equilibrium membrane-bound states the hydrophobic and hydrophilic groups of E5 tend to fall into hydrophobic and hydrophilic lipid environments, respectively. Spatial distributions of hydrophobic/hydrophilic properties were calculated using the MHP approach [23]. The peptide surface was mapped according to the MHP created by its own atoms (MHP$_{peptide}$) and by neighboring atoms of lipids (MHP$_{bilayer}$). This was done for the representative low-energy conformers bound to DMPC and DPPC bilayers (Figure 7). Positive and negative values of MHP correspond to

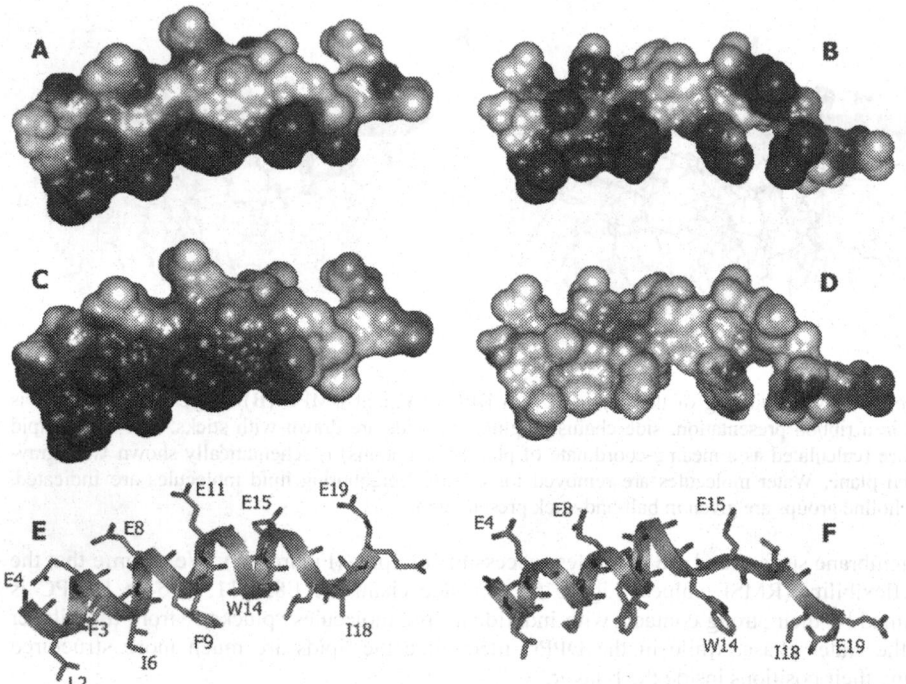

Figure 7. Spatial hydrophobic properties of the peptide surface. The peptide bound to DMPC (**A, C, E**) and DPPC (**B, D, F**) bilayers. The peptide surface is colored according to the values of molecular hydrophobicity potential (MHP) created in the surface points either by the atoms of E5 (**A, B**), or by surrounding lipids (**C, D**). Ribbon representations of the corresponding peptide structures are also shown (**E, F**). Hydrophobic (high MHP) and hydrophilic (low MHP) surface regions are shown in dark and light gray, respectively. Residues revealing strong interactions with lipids are marked.

hydrophobic and hydrophilic surface regions, respectively. The following conclusions can be done. In E5 bound to DPPC the distribution of $MHP_{peptide}$ has two well-defined hydrophobic surface patterns: on the N-terminal helix, and on the C-terminal tail. In the DMPC-bound conformation all these hydrophobic zones form one integral pattern. It is seen (Figure 7A, C, E) that better complementarity of $MHP_{peptide}$ and $MHP_{bilayer}$ distributions is observed for the E5/DMPC complex relative to the E5/DPPC one (Figure 7B, D, F). Better complementarity in the E5/DMPC complex is also observed for polar residues. In the DMPC bilayer unfavorable hydrophobic/hydrophilic contacts are observed for residues G1, I10, and partially for G20. In the DPPC membrane such contacts are realized for L2, F3, A5, I6, A7, F9, I10, L17, and partially for W14, I18 and G20. The hydrophobic residues that fall into nonpolar lipid environment upon binding, demonstrate low energies of van der Waals interactions with the membrane, while the hydrophilic residues involved in favorable contacts participate in strong electrostatic interactions with the membrane. The aforementioned observations confirm the importance of the "hydrophobic match" effects in binding of the fusion peptide. We would like to outline that the results obtained using two independent methods lead to similar conclusions. Moreover, the MHP-mapping complements MD-data. This method can be used for fast delineation of putative membrane binding sites in FPs.

3.2.4. Destabilization of lipid bilayers induced by the fusion peptide

To understand whether the membranes are sensitive to peptide insertion, structural and dynamic characteristics of both "pure" lipid bilayers, as well as of their complexes with E5 were compared based on the analysis of resulting MD trajectories. Within the accuracy of the analysis, the presence of E5 has little effect on the average "global" equilibrium parameters calculated for the whole membranes: the surface area per lipid, the membrane thickness and the order parameters of acyl chains (data not presented). On the other hand, there are prominent differences in distributions of some atomic groups of the lipid/water systems along the membrane normal. Thus, in the E5/DMPC complex the increase of phosphorus density for its upper leaflet as compared with the "pure" bilayer is observed in the outer part of the lipid/water interface (Figure 8A). (Hereinafter this leaflet is called "upper", while the other one is referred to as "lower".) In this case fluidity of the bilayer permits deep immersion of the peptide into the interface. This is accompanied by extrusion of a number of DMPC headgroups toward the water phase (Figure 6B). In contrast, in the E5/DPPC system such a density growth is observed in the inner part of the interface, i.e. the peptide adsorbs from water on the bilayer surface and "pushes" several lipid headgroups toward the acyl chain region of the membrane. We should notice that, according to this criterion in both cases there is practically no effect on the lower leaflets of the bilayers.

Depending on the bilayer content, important differences are also observed in distributions of water along the bilayer normal $(\rho_w(z))$. On the upper interface the peptide induces different changes of the function $\rho_w(z)$ obtained for "pure" bilayers. As seen in Figure 8B, at $z > 0$ the differential curve $\Delta\rho_w(z) = \rho_w(z)_{E5/DPPC} - \rho_w(z)_{DPPC}$ has two negative (at 16 and 22 Å) and one positive (at 11 Å) peaks. Appearance of the negative peak at 22 Å is caused by exclusion of water molecules from the interfacial region by the bound peptide. This is confirmed by the fact that the peptide's density distribution has maximum at 21 Å (Figure 8A). The positive and negative peaks of $\Delta\rho_w(z)$ at 11 and 16 Å are explained by the peptide-induced displacement of a number of lipid headgroups with tightly bound water toward nonpolar core of the bilayer.

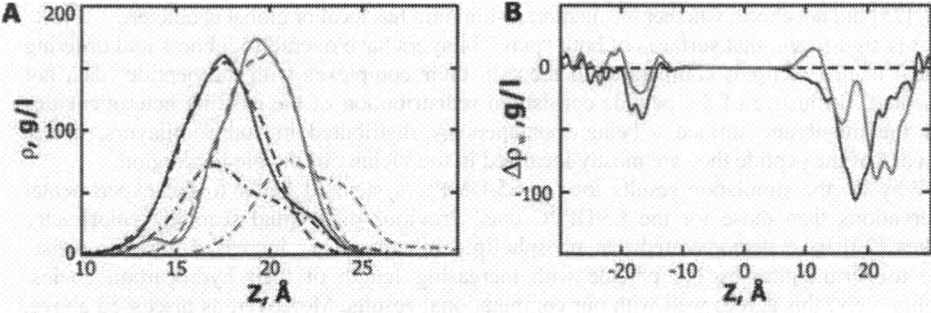

Figure 8. The peptide-induced "membrane response". **A.** Density distributions of phosphorus atoms along the membrane normal for their upper leaflets. MD data in DMPC (*black*) and DPPC (*gray*) bilayers. *Dashed lines*: "pure" bilayers; *solid lines*: bilayers with bound peptide; *chain lines*: the peptide. **B.** Changes in water distributions along the membrane normal $(\rho_w(z))$ caused by the peptide insertion. Differential curves $(\Delta\rho_w(z) = \rho_w(z)_{peptide/bilayer} - \rho_w(z)_{bilayer})$ for E5/DMPC and E5/DPPC systems are shown with *black* and *gray* lines, respectively.

Somewhat different picture in water behavior is observed in the DMPC membrane. In the upper leaflet region the function $\Delta\rho_w(z)$ reveals only one negative peak at 18 Å. Its integral intensity is close to that of the main negative peak (at 22 Å) in DPPC. As before, we attribute this peak to water molecules excluded from the interface by the adsorbed peptide (the peptide's density in the E5/DMPC system has maximum at 17 Å (Figure 8A). Because the left-hand part of the density profile for phosphorus atoms remains unchanged upon the peptide binding (Figure 8B), lipid headgroups are not "squeezed" into the acyl chain region of DMPC – this bilayer is fluid enough to permit their "flow around" the inserted peptide. As a result, water molecules tightly coordinated with the phosphatidylcholine moieties, do not change their distribution along the Z-axis. This explains the absence of additional peaks for the function $\Delta\rho_w(z)$. Interestingly, small (although non-negligible) effects induced by the peptide, are also observed on the lower leaflets of both bilayers: corresponding differential curves $\Delta\rho w(z)$ reveal two positive (at –22 Å and –14 Å) and one negative (at –18 Å) peaks. We attribute these differences to local broadening of the lower water–lipid interfaces.

Additional insight into the microscopic picture of the "membrane response" may be gained via detailed analysis of in-plane distributions of various bilayer characteristics. In-plane distributions of the acyl chains order parameters (S_{CD}) are shown in Figure 9C–D. Areas of low S_{CD} are indicated with gray hatching. It is seen that the peptide causes disordering of lipids in the vicinity of the contact region. Such effects are more pronounced in DMPC, rather then in DPPC bilayer. At the same time, within the errors associated with the technique, we cannot say, whether the peptide influences on the ordering of lipids from lower leaflets of both membranes.

The differential 2D map (Figure 9A–B) pictorially illustrates deviations of the membrane thickness from that obtained for "pure" bilayer. Analysis of this data shows that in the peptide's "shadow" the thickness of DMPC and DPPC bilayers is locally decreased up to 6 and 4 Å, respectively. Similar phenomena were also reported by Vaccaro et al. [25] based on the results of MD simulations of H3 influenza hemagglutinin FP in relatively "loose" POPC bilayer, although in that case the effect was more pronounced: the authors observed a thinning of ~8 Å for the entire bilayer. Taking into account unsaturated character of the POPC molecules, the results of both studies seem to be consistent. However, we should note that Vaccaro et al. [25] did not check, whether the membrane thinning has local or global character.

It is significant, that surfaces of both "pure" bilayers have overall roughness and ordering of acyl chains of lipids comparable to those in their complexes with the peptide (data not presented). Influence of the peptide consists in redistribution of the existing heterogeneities over the membrane surface – being spontaneously distributed in "pure" bilayers, in the presence of the peptide they are mostly localized in the vicinity of the binding region.

Why do the simulation results for the E5/DMPC system fit better to the experimental observations than those for the E5/DPPC one? Previous differential scanning calorimetry studies [30] have demonstrated that phospholipid bilayers show increased intrinsic resistance to perturbation by the peptide with increasing length of their hydrocarbon chains. Qualitatively, this agrees well with our computational results. Moreover, as discussed above, the simulations can be used to propose possible molecular mechanisms of such effects. It will be recalled that the experimental structural information about FPs was mainly obtained in membrane mimics (DPC micelles). So, it seems reasonable to assume that, unlike DPPC, the macroscopic dynamic properties of DMPC bilayer resemble better the micellar environment. Similar conclusion has been reached by Wong et al. [31] based on MD simulations of a fragment of the adrenocorticotropin hormone both, in DPC micelle and in DMPC bilayer.

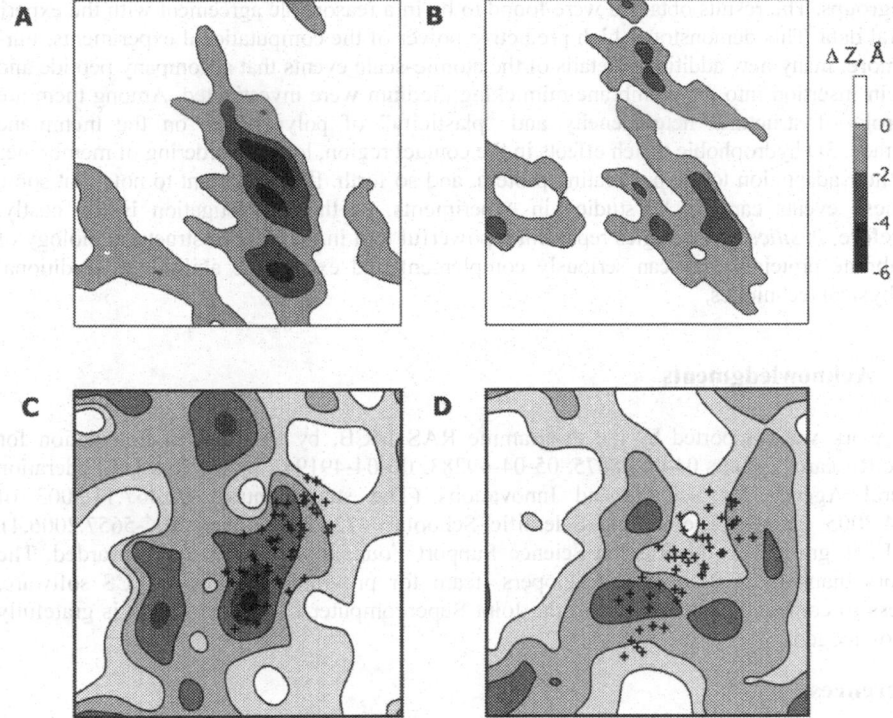

Figure 9. Local destabilization of bilayers. **A–B**. Local thinning (ΔZ) of DMPC (**A**) and DPPC (**B**) bilayers induced by peptide binding. Two-dimensional contour maps obtained via subtraction of the landscapes for upper and lower leaflets of bilayers. The values of ΔZ indicate deviation from the thickness of pure bilayers. They are colored according to the scale shown near the corresponding map. **C–D**. Local disordering of DMPC (**C**) and DPPC (**D**) bilayers induced by the peptide binding. In-plane distribution of order parameters S_{CD} for the acyl chains of lipids. Ordered (S_{CD} ~0.3) and disordered (S_{CD} ~0.0) membrane regions are indicated with light- and dark-gray hatching, respectively.

4.   Conclusions

Nowadays the theoretical methods of simulations of peptides and proteins in explicit lipid bilayers and detergent micelles – with surrounding water molecules and ions – represent quite a realistic approximation for studies of protein-membrane interactions. This class of models conforms reasonably well to available physicochemical data on the structure of such membrane mimics and to processes accompanying binding of peptides and proteins. Here we performed long-term MD simulations of two polypeptides with diverse folds (α-helical and β-structural), different modes of membrane binding, and biological activities. To inspect the role of membrane composition in protein–lipid interactions, the calculations were performed in lipid bilayers and detergent micelles differing in length, chemical nature and/or charge of headgroups. The results obtained were found to be in a reasonable agreement with the experimental data. This demonstrates high predictive power of the computational experiments. Furthermore, many new additional details of the atomic-scale events that accompany peptide and protein insertion into the membrane-mimicking medium were investigated. Among them are the role of structural heterogeneity and "plasticity" of polypeptides on the membrane interface, 3D hydrophobic match effects in the contact region, local disordering of membranes and their adaptation to the penetrating protein, and so forth. It is important to note that some of these events can not be studied in experiments, or their investigation is too costly. Therefore, *in silico* technologies represent a powerful tool in the field of structural biology of membrane proteins; they can seriously complement and extend the abilities of traditional biophysical techniques.

5.   Acknowledgments

This work was supported by the Programme RAS MCB, by the Russian Foundation for Basic Research (grants 04-04-48875, 05-04-49283, 06-04-49194), by the Russian Federation Federal Agency for Science and Innovations (The state contract 02.467.11.3003 of 20.04.2005, grants "The leading Scientific Schools" 4728.2006.4 and MK-5657.2006.4). R.G.E. is grateful to the Russian Science Support Foundation for the grant awarded. The authors thank the GROMACS developers' team for providing the GROMACS software. Access to computational facilities of the Joint Supercomputer Center (Moscow) is gratefully acknowledged.

References

1.   Dahl, S.G., Kristiansen, K., and Sylte, I. (2002). *Ann. Med.* 34, 306.
2.   Yeaman, R.M. and Yount, N.Y. (2003). *Pharmacol. Rev.* 55, 27.
3.   Epand, R.M. (2003). *Biochim. Biophys. Acta* 1614, 116.
4.   Torres, J., Stevens, T.J., and Samso, M. (2003). *Trends Biochem. Sci.* 28, 137.
5.   Forrest, L.R. and Sansom, M.S. (2000). *Curr. Opin. Struct. Biol.* 10, 174.
6.   Roux, B. (2002). *Curr. Opin. Struct. Biol.* 12, 182.
7.   Lazaridis, T. (2003). *Proteins* 52, 176.
8.   Efremov, R.G., Nolde, D.E., Konshina, A.G., Syrtcev, N.P., and Arseniev, A.S. (2004). *Curr. Med. Chem.* 11, 2421.
9.   Gatineau, E., Takechi, M., Bouet, F., Mansuelle, P., Rochat, H., Harvey, A.L., Montenay-Garestier, T., and Menez. A. (1990). *Biochemistry* 29, 6480.
10.   Dubovskii, P.V., Lesovoy, D.M., Dubinnyi, M.A., Konshina, A.G., Utkin, Y.N., Efremov, R.G., and Arseniev, A.S. (2005). *Biochem. J.* 387(3), 807.
11.   Dubovskii, P.V., Dementieva, D.V., Bocharov, E.V., Utkin, Y.N., and Arseniev, A.S. (2001). *J. Mol. Biol.* 305, 137
12.   Dubinnyi, M.A., manuscript in preparation.
13.   Murata, M., Sugahara, Y., Takahashi, S., and Ohnishi, S. (1987). *J. Biochem. (Tokyo)* 102, 957.
14.   Han, X., Bushweller, J.H., Cafiso, D.S., and Tamm, L.K. (2001). *Nat. Struct. Biol.* 8, 715.
15.   Dubovskii, P.V., Li, H., Takahashi, S., Arseniev, A.S., and Akasaka, K. (2000). *Protein Sci.* 9, 786
16.   Mysel, K. and Prizen, L. (1959). *J. Phys. Chem.* 63, 1696.
17.   Lauterwein, J., Bosch, C., Brown, L.R., and Wuthrich K. (1979). *Biochim. Biophys. Acta.* 556, 244.

18. Polyansky, A.A., Volynsky, P.E., Nolde, D.E., Arseniev, A.S., and Efremov, R.G. (2005). *J. Phys. Chem. B* 109(31), 15052.
19. Volynsky, P.E., Polyansky, A.A., Simakov, N.A., Arseniev, A.S., and Efremov, R.G. (2005). *Biochemistry* 44(44), 14626.
20. Efremov, R.G., Volynsky, P.E., Nolde, D.E., Dubovskii, P.V., and Arseniev, A.S. (2002). *Biophys. J.* 83, 144.
21. Lindahl, E., Hess, B., and van der Spoel, D. (2001). *J. Mol. Mod.* 7, 306.
22. Van Gunsteren, W.F. and Berendsen, H.J.C. (1987). *Gromos-87 Manual*. Biomos BV. Nijenborgh 4, 9747 AG Groningen, The Netherlands.
23. Efremov, R.G. and Vergoten, G. (1995). *J. Phys. Chem.* 99(26), 10658.
24. Huang, Q., Chen, C.-L., and Herrmann, A. (2004). *Biophys. J.* 87, 14.
25. Vaccaro, L., Cross, K.J., Kleinjung, J., Straus, S.K., Thomas, D.J., Wharton, S.A., Skehel, J.J., and Fraternali, F. (2005). *Biophys. J.* 88, 25.
26. Nagle, J.F. and Tristram-Nagle, S. (2000). *Curr. Opin. Struct. Biol.* 10, 474.
27. White, S.H. and Wimley, W.C. (1999). *Annu. Rev. Biophys. Biomol. Struct.* 28, 319.
28. Killian, J.A. (1998). *Biochim. Biophys. Acta* 1376, 401.
29. Webb, R.J., East, J.M., Sharma, R.P., and Lee, A.G. (1998). *Biochemistry* 37, 673.
30. Surewicz, W.K. and Epand, R.M. (1986). *Biochim. Biophys. Acta* 856, 290.
31. Wong, T.C. and Kamath, S. (2002). *J. Biomol. Struct. Dyn.* 20, 39.

ANYTHING GOES – PROTEIN STRUCTURAL POLYMORPHISM

ANGELA M. GRONENBORN
Department of Structural Biology
School of Medicine, University of Pittsburgh
Pittsburgh, PA 15261, USA
*Corresponding author: amg100@pitt.edu

Abstract: Immunoglobulin binding domain B1 of streptococcal protein G (GB1), a small (56 residues), stable, single domain protein, is one of the most extensively used model systems in the area of protein folding and design. Variants derived from a library of randomized hydrophobic core residues revealed alternative folds, namely a completely intertwined tetramer, a domain-swapped dimer, and a side-by-side dimer. Single amino acid changes are responsible for inducing these different oligomeric structures, frequently involving major structural rearrangements of the wild–type-like monomer structure.

1. Introduction

Evolution over billions of years has created and refined proteins for a vast variety of purposes. In many respects nature's intricately designed shapes remain unsurpassed with respect to the level of control encoded at the amino acid sequence level. The ability to produce such extraordinarily complex molecules with virtually error-free positioning of each constituent atom is one of the most striking capabilities of living organisms. Looking at, and describing three-dimensional protein structures we hope to gain insight into evolution and function, but their diversity renders these molecules equally seductive and daunting. Given that we cannot yet assign a structure to a given amino acid sequence with any reliability, although Anfinsen's remarkable discovery that the sequence itself contains the necessary information for folding was made almost 50 years ago, how can we hope to design such molecules for particular function? One promising approach is to follow nature's lead by generating diversity through random mutation, trading the presently intractable deterministic challenge of designing an "improved" molecule for the (possibly) more tractable probabilistic challenge of designing an ensemble likely to contain such a protein. Such ensembles are contained in protein libraries, popular tools that are used increasingly in protein engineering. Protein libraries can be comprehensive, with all twenty amino acids at every position, or focused, with variants differing at few positions containing a subset of all possible amino acids.

Several biological approaches have been developed to increase the likelihood of obtaining "good" proteins in random libraries and the search for functional and improved proteins is necessarily guided by the search for sequences encoding folded proteins. Frequently, – though

J. D. Puglisi (ed.), Structure and Biophysics – New Technologies for Current Challenges in Biology and Beyond, 41–47.

not always, – amino acid types or positions are chosen based on native structures. In the most general sense, very little to nothing is known about how sequence changes affect function, except that (1) most loss of function mutants disrupt the structure and (2) mutations are necessary to improve or change function. It therefore is mandatory to understand the tradeoff between sequence diversity and folding. Consequently, stringent selection schemes are normally imposed when using libraries, limiting the entire sequence, and structure space to those shapes that are known to carry out particular functions. As a result, alternative folded states have only rarely been described for natural proteins.

Despite of the rapid increase in the number of 3D structures of multimeric proteins, the underlying mechanism of oligomerization is largely unknown. Based on all the presently available data, it seems compelling to invoke the involvement of partially folded species, given their strong propensity for aggregation. As pointed out by Perutz in an insightful note (1997), mutations of "internal" residues might result in a loss in free energy of stabilization which, even if small, might lead to a disruptive, "loosening" effect on the native structure. If the resulting structure would consist of a mixture of secondary structure elements, interspersed with flexible, coil-like regions, collision of two molecules would allow insertion of elements belonging to one chain into the other chain. In this manner oligomers or polymers of polypeptide chains could built up. Although, at present, no direct experimental proof exists that in amyloid fibers individual protein molecules are associated via domain exchange, for numerous proteins strong circumstantial evidence lends credence to this mechanism.

2.   Structures

Over the last decade, we have employed the immunoglobulin-binding domain B1 of strepto-coccal protein G (GB1) as our model system for studying structure, dynamics, folding, and oligomerization. GB1 is a small (56 residues), stable, single-domain protein with one α-helix and a four-stranded β-sheet. We created a library of hydrophobic core mutants of GB1 in which nine positions were randomly permuted to comprise the amino acids Leu, Val, Ile, Phe, and Met (Gronenborn et al., 1996). Among these mutants, intriguingly, several oligomeric species were detected and 3D structures for two of these, an extensively intertwined tetramer and a domain-swapped dimer were determined (Byeon et al., 2003; Frank et al., 2002). The latter is a single-site revertant (L5V/F30V/Y33F/A34F) of the quintuple tetramer mutant (L5V/A26F/F30V/Y33F/A34F), thus a single amino acid change transforms the tetramer into the domain-swapped dimer (Figure 1). Interestingly, fibril formation was readily achieved for this mutant and other variants that can exist as domain-swapped dimer structures (Louis et al., 2005). In contrast, no fibrils were obtained under the same experimental conditions with another single site mutant (L5V/F30V/Y33F) that folds into the stable monomer wild-type GB1 fold or for a variant that comprises a highly destabilized, fluctuating ensemble of random, folded, and partially folded structures. These observations imply that a specific intermediate may be required for amyloid formation.

Comparison of the domain-swapped dimer structure with the wild-type monomer suggested Phe34 as a pivotal side chain for the switch from the monomer to dimer, since reverting the sequence back to the wild-type Ala residue yielded a wild-type-like monomer protein again. We, therefore, suspected that a single residue mutant, Ala-34 to Phe-34 in the wild-type sequence context may also induce a similar change in quaternary structure. Intriguingly, the nuclear magnetic resonance (NMR) structure for GB1^{A34F} revealed that this mutant does not exhibit domain-swapping but forms another dimer, namely a side-by-side dimer.

2.1.   COMPARISON BETWEEN GB1 VARIANTS

The quaternary structures of GB1 variants to date are illustrated in Figure 1. So far five different quaternary architectures have been observed experimentally in high resolution structures: the wild-type monomer fold (Figure 1D), a solution side-by-side dimer (Figure 1A), a domain-swapped dimer (Figure 1B), a crystallographic side-by-side dimer (Figure 1C), and a completely intertwined tetramer (Figure 1E). In the solution side-by-side dimer of GB1, the monomer unit is similar to that of the wild-type monomer and the crystallographic side-by-side dimer.

Although the monomer unit of the GB1 side-by-side dimer is similar to wild-type GB1, two distinct regions are involved in the dimer interface. The first interface is formed by the anti-parallel β-sheet which is created between the two β2 strands, one from each polypeptide chain. The second interface includes residues in the helix and loop-I, with Tyr-33 playing a key role. Its aromatic ring swings out from one monomer and inserts snugly into a pocket lined by the side chains of Leu-12 and Asn-37 in the other monomer.

The major difference between the solution side-by-side dimer and the arrangement observed for GB1 in the crystal lattice consists of a flip in the orientation of the entire molecule. In the side-by-side-dimer, both α-helixes lie on the same side of the contiguous β-sheet, whereas in the crystallographic dimer, the monomer has the helix on top and the other at the bottom. Naturally, this type of arrangement is simply a reflection of the crystal symmetry, which does not result in a "true" eight-stranded β-sheet.

In the domain-swapped dimer exchange of the β3-β4 hairpin between the subunits occurs, resulting in alternation between monomer chains for adjacent hairpins in the eight-stranded β-sheet. Half of the dimer, composed of the first β-hairpin and the α-helix (residues 1–37) from one polypeptide chain and the second β-hairpin (residues 42–56) from the other chain, is essentially identical to the monomer structure with a pair-wise r.m.s.d. value of 1.5 Å for the backbone atoms. The "open interface" resides between two β2 strands, one from each polypeptide, and the "closed interface" is found between β1 and β4. In addition to the newly created "open interface" in the domain-swapped dimer, further intermolecular contacts exist between the two α-helixes and their proceeding loops. The helixes are arranged in an antiparallel manner, crossing at their C-termini. This results in a wide V-shaped appearance. These V-shaped α-helixes are buttressing the β-sheet to form a tightly packed hydrophobic core. The only residues for which significant difference in the backbone φ and ψ torsion angles between the wild type and the mutant are observed are those in the linker connecting

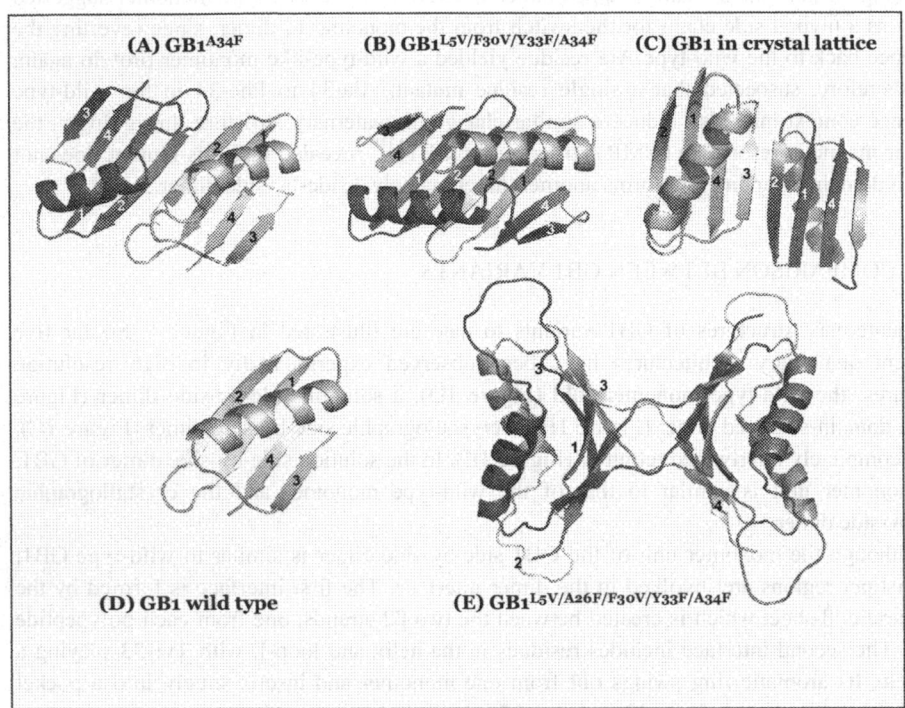

Figure 1. Comparison between different GB1 structures. Ribbon diagrams of wild-type GB1, the domain-swapped dimeric mutant, the side-by-side dimeric mutant, the HS#124 tetramer and the dimer in the crystal lattice are shown. The structures of wild-type GB1 and the side-by-side dimer are aligned based on the inertia tensor of the side-by-side dimer structure. The tetramer #HS124 is scaled to fit into the Figure. In each structure, the first monomer unit is colored green, the second unit blue, and the third and fourth units in tetramer are colored pink and orange, respectively. β-strands are labeled 1, 2, 3, and 4, the α-helix is marked α. In #HS124, only the first unit is labeled for clarity.

the α-helix to strand β3. In contrast to most domain-swapped protein structures reported to date in which the structural elements that are involved in the swap are single units from either the N- or C-terminus of the protein, the present structure has essentially half of the molecule swapped.

An overall comparison between the GB1 side-by-side dimer, wild-type GB1 and the domain-swapped dimer by superimposing the β-strands reveals that the β-sheet regions are very similar, whereas clear differences are observed for the helix and the proceeding loop. In that regard it is intriguing that in the domain-swapped dimer, the helix is preserved, although 4 residues, including Phe-34 are mutated in the sequence, whereas in the side-by-side dimer, containing only the Phe-34 mutation, parts of the helix are dissolved. Comparing the domain-swapped dimer with the GB1 side-by side dimer shows a significant change in angle for the helix; it is tilted by ~17 degree from that observed in wild-type GB1 and the GB1 side-by-side dimer. Interestingly, the reoriented helix in the domain-swapped dimer allows for the side chain of Phe-33 to pack within the hydrophobic core without clashing with Phe-34, clearly different from what is observed for the GB1 side-by-side dimer.

In contrast to clearly discernable structural resemblances between the monomer and the different dimeric structures, hardly any similarities with the tetramer are apparent. In the tetramer structure all interactions between β-strands are intermolecular, and each pseudo-dimeric half contains a six-stranded β-sheet. The β-sheet in the domain swapped dimer is a regular, contiguous eight-stranded sheet, with two hairpin strands alternating between individual polypeptides. Thus, the arrangement of the β-stands in the swapped dimer structure is β3, β4, β1', β2', β2, β1, β4', β3', while in the tetramer one of the two sheets comprise strands β3, β4'', β1', β1, β4''', β3', with an equivalent arrangement (β3'', β4 β1''', β1'', β4', β3''') in the other half. In addition, a shift in register by two amino acids is observed for the β3, β4' sheet in the tetramer. The pseudo-dimeric half of the tetramer is less tightly packed than the side-by-side or swapped dimer structures; this is a natural consequence of the loss of two β-stands.

2.2. DIMER-MONOMER EQUILIBRIUM

In the original ^{15}N-^{1}H HSQC spectra of both, the domain-swapped dimer and the side-by-side dimer, a small number of additional resonances were observed. Upon dilution, these resonances grow in intensity while several resonances clearly assigned to the dimeric species disappeared. The new set of resonances arise from a monomeric form and the monomer-dimer equilibria for both mutants were confirmed using size-exclusion chromatography in conjunction with in-line multiangle light scattering and refractive index detection.

Using 2D NMR exchange experiments at an intermediate concentration where both set of resonances are of comparable intensities, cross peaks between both forms can be observed, confirming that exchange occurs on the microsecond to millisecond timescale.

The dissociation constants (K_d) for the GB1 side-by-side dimer and the domain swapped dimer were calculated from resonance intensities in the ^{15}N-^{1}H HSQC spectra for the same residue in the dimer and monomer species, respectively. The resulting values were ~30 μM and ~ 90 μM. Interestingly, multiangle light scattering with refractive index detection revealed two well-separated peaks for the domain-swapped dimer, with the first and second peaks corresponding to dimeric and monomeric molecular mass species, respectively. In contrast, for the side-by-side dimer a single peak was observed, with a refractive index scattering profile across the peak decreasing from the apparent molecular mass of a dimer down to that of a monomer. Therefore, exchange is fast on the gel-filtration timescale (min), indicative of a smaller kinetic barrier separating the side-by-side dimer from its monomeric conterpart compared to the domain-swapped dimer equilibrium

2.3. WHAT IS THE STRUCTURE OF THE MONOMERIC FORMS IN THESE EXCHANGING SYSTEMS?

In contrast to the domain-swapped dimer and intertwined tetramer, formation of a side-by-side dimer does not necessitate a major structural rearrangement of a wild-type-like monomer structure. For the tetramer, we only observed unfolded monomer and folded tetramer, without any indication of an intermediate species. For both dimers, monomeric folding intermediates can be observed, which become the predominant species at low concentration. Structural characterization of these monomeric species revealed that they exhibit extensive conformational heterogeneity for a substantial portion of the polypeptide chain. Exchange between the conformers within the monomer ensembles on the micro- to millisecond timescale renders the majority of backbone amide resonances broadened beyond detection. Despite these extensive temporal and spatial fluctuations, the overall architecture of the monomeric species resembles that of wild-type GB1, although the protein is only partially folded, exhibiting substantial structural plasticity.

3.   Conclusion

Several quite distinct quaternary arrangements for the small protein GB1 have been observed. Their amino acid sequences are very similar to wild type and were derived from a library of random core mutants.

It is tempting to speculate that structural plasticity in a partially destabilized monomer unit constitutes the prerequisite for oligomerization, with distinct partially folded intermediates being pivotal for the formation of the individual architectures. In this sense, ordered polymerization, as observed in amyloid fibrils, may also require specific destabilized monomeric species. Our studies using a core sequence library of the GB1 fold provide ample experimental evidence for this notion. Given that a great deal of biology arises from the transient existence of protein conformations of only limited stability, particularly important in signaling networks, and in the formation of protein complexes, understanding the nature and requirements for structural plasticity is of immense importance. Therefore, our data on the different oligomers of GB1 should supply valuable experimental data for testing different theoretical models of protein folding, design, and evolution.

4.   Acknowledgements

I am indebted to all members of my research group, past and present, who contributed to the entire "GB1 story". In particular, In-Ja Byeon, Kirsten Frank, Jungoo Jee, and John Louis played important roles. In addition, technical support by J. Barber and M. Delk is gratefully acknowledged.

References

1. Byeon, I.J., Louis, J.M., and Gronenborn, A.M. (2003). A protein contortionist: core mutations of GB1 that induce dimerization and domain swapping. *J. Mol. Biol.* 333, 141–152.
2. Byeon, I-J.L., Louis, J.M. and Gronenborn, A.M. (2004). A captured folding intermediate involved in dimerization and domain-swapping of GB1. *J. Mol. Biol.* 340, 615–625.
3. Frank, K.M., Dyda, F., Dobrodumov, A., and Gronenborn, A.M. (2002). Core mutations switch monomeric protein GB1 into an intertwined tetramer. *Nat. Struct. Biol.* 9, 877–885.
4. Gronenborn, A.M., Frank, M.K., and Clore, G.M. (1996). Core mutants of the immunoglobulin binding domain of streptococcal protein G: stability and structural integrity. *FEBS Lett.* 398, 312–316.
5. Louis, J.M, Byeon, I-J.L., Baxa, U., and Gronenborn, A.M. (2005). The GB1 amyloid fibril – Recruitment of the peripheral β-strands of the domain-swapped dimer into the polymeric interface. *J. Mol. Biol.* 348, 687–698.
6. Perutz, M.F. (1997). Mutations make enzyme polymerize. *Nature,* 385, 773–775.

TANDEM INTERACTIONS IN THE TRP REPRESSOR SYSTEM MAY REGULATE BINDING TO OPERATOR DNA

OLEG JARDETZKY AND MICHAEL D. FINUCANE
Department of Chemical and Systems Biology
Stanford University, Stanford, CA, USA

To whom correspondence should be addressed:
Department of Chemical and Systems Biology
Stanford University, Stanford, CA 94305-5174, USA
Telephone 650 723-6153
Fax 650 723-2253

Abstract: Recent SPR studies illustrate that the AV77 super-repressor mutant of TR acts by considerably lengthening the lifetime of the protein–DNA complex, and that this is achieved by enhancing the population of a particularly stable form of the complex. Acting on the hypothesis that this population represents a tandem complex of the protein with DNA, we designed a mutant to enhance protein–protein interactions. The resulting mutant, SC5, acts as a super-repressor. Based on this information, we propose that the tandem interaction is crucial to the formation of long-lived complexes on DNA. An examination of the crystal structure of the tandem complex leads us to suggest that one role of tryptophan may be to enhance the tandem interaction of TR with its operator DNA. The hypothesis also provides a role for the transition of the DNA-binding helices from being flexible to being more rigid when the corepressor is bound, and it may also explain why mutations at position 77 are often super-repressors.

Keywords: *trp* repressor; tandem binding; super-repressor; allostery; DNA recognition

Abbreviations: SPR: Surface Plasmon Resonance; RU: resonance units; TR: tryptophan repressor; WT: wild-type

1. Introduction

Tryptophan repressor (TR) is a dimeric protein which controls the transcription of genes that code for tryptophan production and for the repressor itself. The operators that bind TR are *aroH*, *aroL*, *trpEDCBA*, *mtr*, and *trpR*. Although these operators have a high degree of homology they are not identical. Apart from minor differences in recognition sequence, the operators are able to bind different numbers of TR dimers, depending on the length and symmetry of the operator sequences. For example, *trpEDCBA* can bind three dimers, *AroH* can bind two dimers, but *trpR* only binds a single TR dimer.[1] In 1993, the crystal structure of a 2:1 tandem complex was published[2] showing that indeed, several dimers could pack neatly together on the DNA, wrapping around the DNA. These results and others led to the suggestion that the ability of TR to bind in tandem could provide a higher degree of control in the system.

J. D. Puglisi (ed.), Structure and Biophysics – New Technologies for Current Challenges in Biology and Beyond, 49–64.

Early studies showed[3] that the binding of holoTR to the *trpR* operator was much weaker than that to the *trpEDCBA* operator, so much so that the complex was not stable enough to be visible in gel retardation studies. When we compared binding of WT TR to both operators using surface plasmon resonance (SPR), we found that the complex with *trpR* dissociates rapidly ($t_{1/2} \sim 4s$) and completely in a single step, whereas the complex with *trpEDCBA* dissociates in two distinct phases. Almost 60–70% of the TR dissociates rapidly, but there is a smaller population which remains and dissociates at a much slower rate.[4] The most significant difference between the complexes is that the *trpEDCBA* operator allows for tandem binding of several repressors to the DNA, whereas the *trpR* operator only possesses a single canonical-binding site.[1] It is therefore tempting to suggest that the kinetically distinct populations represent 1:1 and tandem repressor–DNA complexes, and that tandem binding acts to greatly increase the lifetime of the operator–repressor complex.

In order to test this hypothesis, we designed a mutant to enhance the protein–protein interaction, to see if the population of the slowly dissociating population could be increased. The method we chose was to introduce a cysteine which could form an intermolecular disulfide bond. Fortunately, cysteine is not part of the WT sequence as this could allow the formation of a number of alternative disulfides, complicating the result. The advantage of choosing the disulfide to crosslink the dimers, is that this is a reversible covalent bond. It is stable enough not to complicate the kinetics yet can be broken experimentally by adding a reducing agent. We decided to introduce the cysteine at position 5 on the N-terminal arm. In the 1:1 complex with operator DNA, the crystal[5] and NMR[6] structures show that the n-terminal arm of TR is unstructured which should allow the cysteine to react with the cysteine on an incoming TR dimer. Serine 5 was chosen as the point of mutation, as the mutation is a semi-conservative one, replacing an oxygen atom with a sulfur.

If our hypothesis was correct, this mutation would be expected to substantially enhance the population of the slow-dissociating complex, resulting in a super-repressor (defined operationally here as a mutant with enhanced binding characteristics), and the effect would be expected to be easily reversible on addition of a reducing agent. Our experiments confirmed that these expectations are met, supporting the hypothesis.

2.   Results and Discussion

TR was bound in the presence of tryptophan-containing buffer to biotinylated operator DNA immobilized on a streptavidin chip surface, as previously characterized.[4] We define the term R_A as the amount of repressor which has bound during the association phase (e.g. the response in resonance units (RU), at $t = 1000s$ in Figure 1). After 750 s of binding, the chip surface was washed for 1,000 s with trp-containing buffer, resulting in a loss of TR from the surface. We define R_D as the amount which remains bound at the end of the dissociation phase (e.g. the response in RU, at $t = 2000s$ in Figure 1).

Four mutants were tested, as well as the WT. All five proteins bound rapidly to the surface, but WT, AL77, and AT77 reached >90% of R_A within 30 s of binding, whereas SC5 and AV77 had only reached about 25% and 50% of their respective R_A values within this time (Figure 1). After this initial period, the mutants SC5 and AV77 continued to associate at a slower rate with the 20-base pair operator sequence (*20-consensus*, see Table 1 for this and all operator DNA sequences used). Similar divergence in the behavior of the proteins occurred

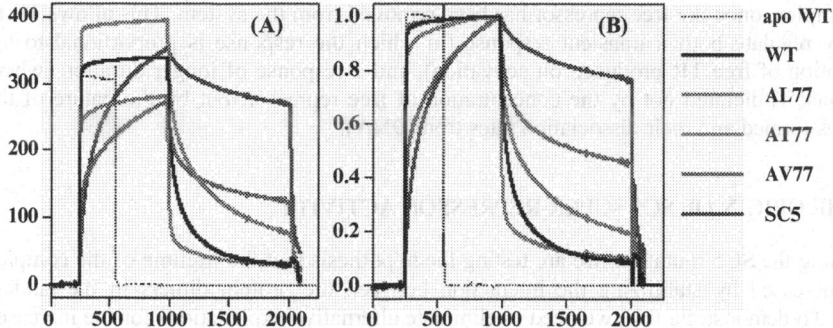

Figure 1. SPR traces of TR binding to a 20-base pair consensus operator sequence (see Table 1 for the sequence). The binding profiles for WT and AV77 were determined previously,[4] the results for AL77, AT77 and SC5 are from the present study. (**A**) Raw data (baseline corrected by subtraction of data from an underivatized streptavidin surface), (**B**) Data from (**A**) normalized such that $R_A = 1$.

during the dissociation phase. In contrast to the WT aporepressor which dissociates in a single rapid step (Figure 1), all five holorepressors have a slowly dissociating component.

TABLE 1. Operator DNA sequences studied

(1) 5' b-AAAAAT**GTACTA**GTTAACT**AGTACA**   *20-consensus*
            A**CATGA**TCAATTGA**TCATGT**

(2) 5' b-AAAAATTAATCATC**GAACTA**GTTAACT**AGTAC**GCAAGTTCACGTA   *40-trpEDCBA*
            AATTAGTAG**CTTGA**TCAATTGA**TCATG**CGTTCAAGTGCAT

(3) 5' b-AAAAAGATATGCTATC**GTACT**CTTTAGCG**AGTAC**AACCGGGGGAG   *40-trpR*
            CTATACGATAG**CATGA**GAAATCGC**TCATG**TTGGCCCCCTC

It is possible to rank the complexes according to the population of the tightly bound complex. Apo WT TR/*20-consensus* and holo WT TR/*40-trpR* have no significant population which remains bound after 30 s. The complexes of *20-consensus* with holorepressors rank in order of increasing population of the tightly bound complex (as measured by R_D): WT,AL77 < AT77 < AV77 << SC5.

1,000 s into the dissociation phase, each repressor still has a fraction bound to the operator. For WT and AL77, ~10% remains. For AT77, AV77, and SC5, the fraction remaining bound is ~20%, ~45%, and ~75%, respectively (Figure 1). The mutant SC5 therefore forms a substantially more long-lived complex with consensus operator than even the hinge mutants AL77, AT77, and AV77. What seems clear from the experiments, is that mutants AV77 and SC5 are not super-repressors because of enhanced association rates with the DNA (these appear to be slower), but rather because the dissociation rates have been decreased. Furthermore, it is clear that there are at least two forms of the complex in each case; one which breaks down rapidly, and a second which is considerably longer-lived. Because the binding and dissociation do not follow simple 1:1 langmuir binding kinetics, there are two separate ways in which the strength of the repressor–operator association may be defined; (1) the relationship between the protein concentration and the amount of complex which is formed, which in 1:1 kinetics could be expressed as a binding constant and (2) the longevity of the complex, once the free repressor has been removed from the system. This allows TR to separately regulate both a transient response (in which the response is proportional to the concentration of free TR produced on activation), and a response of longer duration (where the response is dictated not by the concentration of free repressor, but by the nature of the complexes formed and their dissociation rates from DNA).

2.1. THE ORIGIN OF SC5 SUPER-REPRESSOR ACTIVITY

In designing the SC5 mutation, we are testing the hypothesis that the lifetime of the complex can be increased by stabilizing the interaction between the protein dimers in the tandem complex. To demonstrate this, we need to eliminate alternative explanations for the increased binding of SC5 to DNA.

SC5 could stabilize the complex in either of three ways; (1) through a specific noncovalent interaction between the cysteine side chain and the DNA, (2) via a Michael addition to a base pair (e.g.[7]), or (3) through crosslinking of protein dimers with a disulfide bond. To distinguish between these possibilities, we associated both AV77 and SC5 TR with 20-consensus operator in the presence of 2 mM trp, and then washed with three different buffers (the protein is reloaded prior to each wash, Figure 2). The first buffer contained no tryptophan, and the second contained 2 mM tryptophan, testing the sensitivity of the binding to ligand. The third contained 2 mM tryptophan and also 10 mM dithiothreitol (DTT).

Both AV77 and SC5 acted as repressors in the presence of tryptophan, but washed off rapidly in its absence (Figure 2). DTT abolished binding of SC5 to the operator, but had little effect on AV77 (Figure 2), indicating that while both mutants require tryptophan for activation, only SC5 also requires oxidising (ambient) conditions. In all cases, the surface was regenerated successfully by flushing the cells with salt (see methods) indicating that the binding remains salt-sensitive.

If (1) the decrease in dissociation rate were due to an interaction between the DNA and the cysteine side chain in its reduced, unmodified form, then the addition of DTT should have little effect on the dissociation rate of the complex. The possibility that DTT interferes in some other manner with the interaction between protein and DNA has been eliminated, as the AV77 mutant shows no comparable effect. These results demonstrate unambiguously that some reducible and reversible modification of the cysteine side chain is responsible for the increased repression, and not a noncovalent interaction due to the conservative (oxygen → sulfur) mutation at the interfacial region. If the increased repression were due to noncovalent interactions, then the presence or absence of DTT would have little or no effect, as with AV77.

The formation of a Michael adduct (2) can also be ruled out, because it is unlikely to be reversed by DTT in the manner observed, and more clearly, such an adduct would not be broken down by the removal of tryptophan, or by the addition of 1 M salt, as is observed (Figure 3). After eliminating both of these possibilities, we are left with (3), the formation of an interprotein disulphide. Although the equivalent serine residues are not close to each other in the tandem complex (PDB code 1TRR[2]), they are in a highly flexible region of the molecule, which is unstructured in the 1:1 complex with DNA as probed by both X-ray crystallography and NMR.[5,6] Fitting the curves (Figure 3) to the data for this model (Scheme 1) resulted in acceptable fits at low protein concentrations, although we note (1) that other models also yield acceptable fits, and (2) that at higher concentrations, fits to this model were not acceptable (presumably for the same reasons – nonspecific binding, aggregation – as seen for the WT[4]. It is quite probable that the model needs to be expanded, to account for (1) more than two dimers binding in tandem (2) nonspecific binding at higher protein concentrations (3) possible extra steps such as conformational changes, or secondary binding modes. The introduction of such extra factors would however lead to an unacceptable increase in the number of free parameters. Nevertheless, the data are in reasonable agreement with our model.

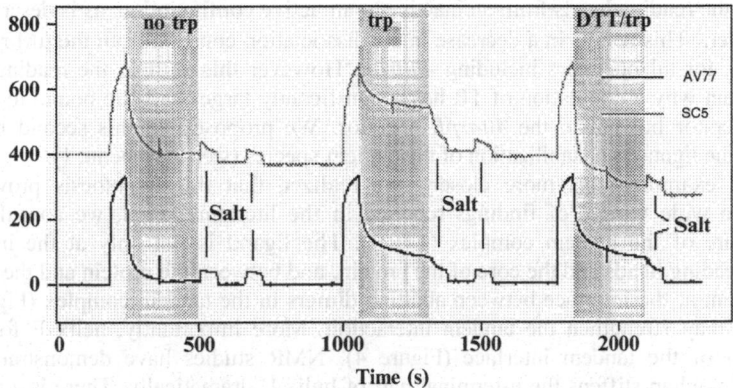

Figure 2. SPR traces of AV77 and SC5. In each case, the repressor is bound in the presence of 2 mM tryptophan. The dissociation phases are shaded and indicate the use of buffers without tryptophan, with 2 mM tryptophan, and with both 2 mM tryptophan and 10 mM DTT, respectively, as indicated. The traces have been referenced to a lane which contained no DNA bound to the chip surface.

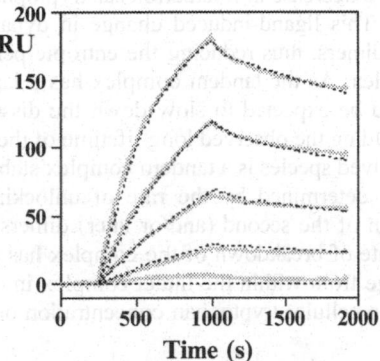

Figure 3. SPR traces of the binding of holo SC5 to; and the dissociation from, the *20-consensus* operator, at concentrations of 0, 2, 5, 10, and 20 nM (protein). The *solid gray lines* are the experimental data, while the dashed lines are the fitted lines ($k_{a1} = 1 \times 10^8$ M^{-1}s^{-1}, $k_{d1} = 0.22$ s^{-1}, $k_{a2} = 6.9 \times 10^5$ M^{-1}s^{-1}, $k_{d2} = 2.6 \times 10^{-3}$ s^{-1}, $R_{max} = 200$ RU, k_t (mass transport coefficient) $= 2.3 \times 10^7$. $\chi^2 = 7.65$.

2.2.   A NEW ROLE FOR LIGAND BINDING IN THE ACTIVATION OF TR

The existence of a stable population in the holoWT TR interaction with *20-consensus* operator DNA (which allows for tandem binding) and the absence of such a stable population in the absence of tryptophan, suggested to us that a major role for the ligand might be to stabilize the tandem interaction between repressor dimers in the complex. It is important at this juncture to distinguish two forms of activation of TR. The first form of activation is the movement of the reading heads from an inactive to an active configuration, as is described by Luisi and Sigler.[8] This results in a decrease in the dissociation constant from the µM range to the nM range, for all operators including *40-trpR*. However this shift in the reading heads does not explain why a population of TR has a significantly larger half-life bound to the *20-consensus* operator but not to the *40-trpR* operator. We propose that this second form of activation by the ligand is a stabilization of the tandem species (step 2, Scheme 1).

When we examined this more closely, we realized that this hypothesis provided a rationale for a wide variety of findings reported in the literature. First, we consulted the crystal structure of the tandem complex (1TRR). The ligand is not only at the interface between the reading heads and the core of the protein, and between the protein and the DNA;[8] it is also present at the interface between abutting dimers in the tandem complex (Figure 4). This in itself may strengthen the tandem interaction. More importantly, helix E forms an extensive part of the tandem interface (Figure 4). NMR studies have demonstrated that binding of tryptophan stiffens the n-terminal half of helix E dramatically. There is a definite increase in the number and strength of the intrahelical NOEs in this region as tryptophan is bound, yielding a more precisely defined structure for the holorepressor than for the aporepressor.[9] Furthermore, the solvent exchange rates reveal a definite slowing when tryptophan is bound, beyond that which can be ascribed to the solvent exclusion that takes place as the ligand is bound.[10] These experimental findings are supported by molecular dynamics studies[11] which show a decrease in fluctuations as tryptophan binds to the repressor, in the D & E helical regions. This ligand-induced change in dynamics, may pre-order the nascent interface between the dimers, thus reducing the entropic penalty associated with the formation of the tandem complex. As the tandem complex has a larger interface with DNA than a single dimer, this would be expected to slow down the dissociation of the repressor from DNA (k_{d2}, Scheme 1), yielding the observed long lifetime of the complex.

If, as we suggest, the long-lived species is a tandem complex stabilized by ligand, then the breakdown of the complex is determined by the rate of unlocking of the complex (k_{d2}, Scheme 1), and the dissociation of the second (and/or later) dimers. However this raises an important point; because the rate of breakdown of the complex has a very long half-life, the ligand must be able to exchange from within the intact complex in order for the repressor to sense and respond to changes in cellular tryptophan concentration on a reasonable timescale.

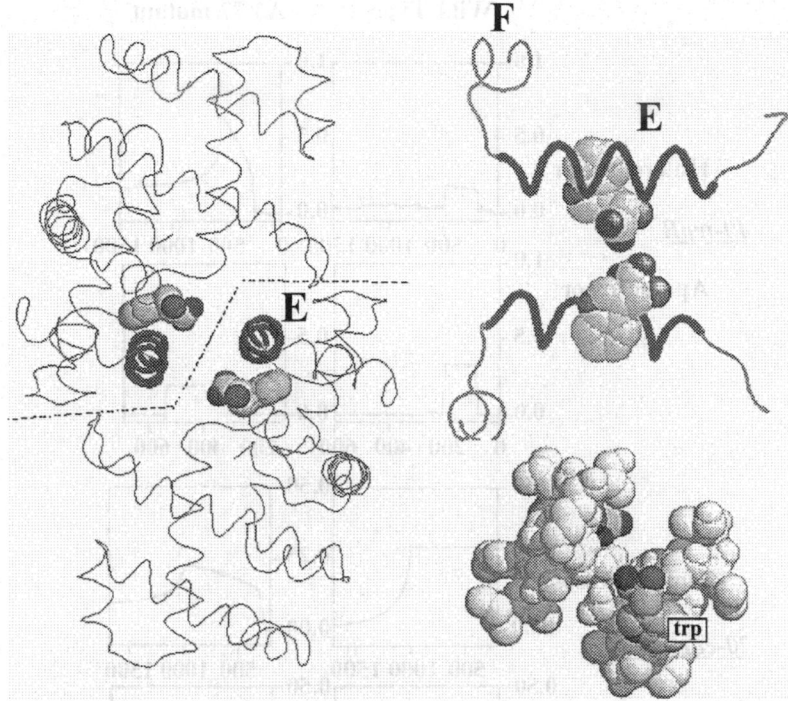

Figure 4. The tandem complex of TR with operator DNA (PDB: 1TRR). For clarity, the DNA has been omitted from the figure. The tandem interface between adjacent dimers is indicated with a dashed line (*left*). Two views of the interplay of Helix E and the ligand tryptophan at the dimer interface are presented on the *right*. The top indicates the close association of the ligand with the helix, while the lower shows the close association between the adjacent dimers.

NMR measurements measure the k_{off} for tryptophan in the WT complex to be ~3.5s,[12] demonstrating that it does indeed exchange from within the complex as would be required. It would therefore appear likely that although the ligand is required for stabilization of the long-lived complex (Figures 1 and 2), its loss from the complex may not mandate immediate dissociation of the complex. Indeed, comparison of the rate of decay of the holo AV77–20-*consensus* complex once tryptophan has been removed from the buffer (Figure 2) with the decay of the apo AV77 complex (Figure 5) would seem to suggest that there is a lag time between the loss of the ligand and the breakdown and dissociation of the complex, at least for AV77.

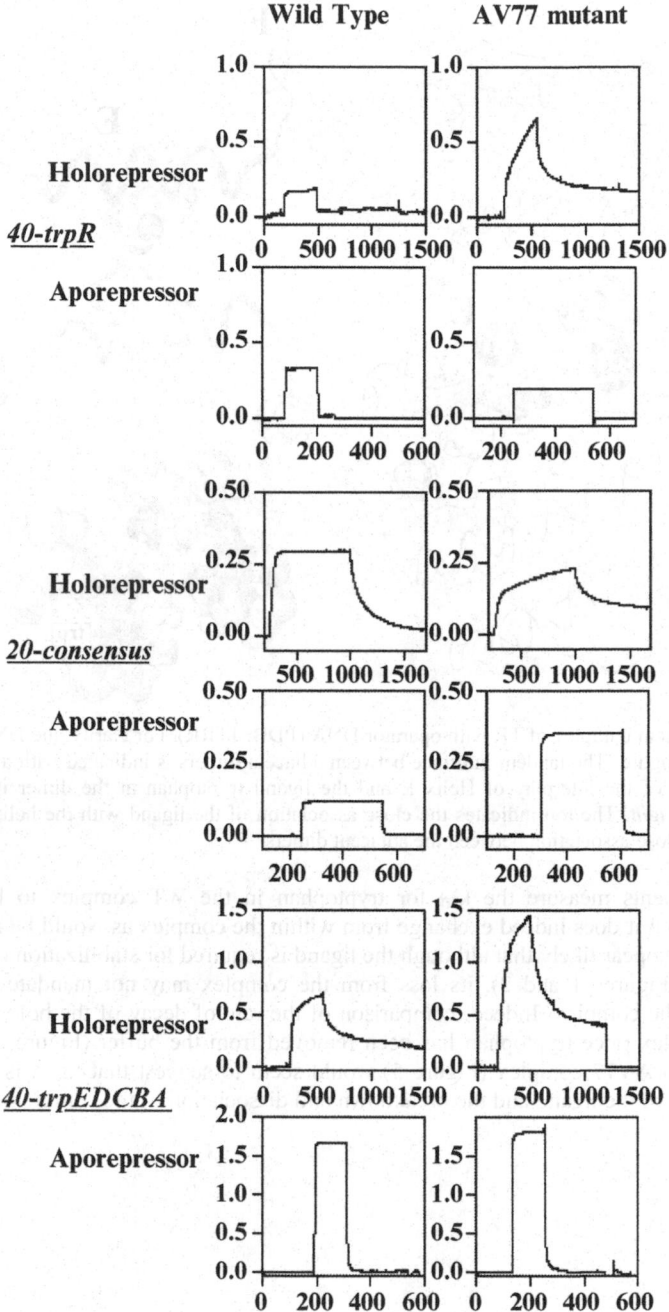

2.3. THE SUPER-REPRESSOR AV77

AV77 has been shown to be a super-repressor in vivo,[13,14] but contradictory results resulted in confusion as to whether the binding constant was the same as,[15,16] or different than[17,18] that of the WT. Using SPR, we confirmed the results of Marmorstein[17] which showed the K_D of AV77 aporepressor to be approximately 8× less than that of the WT aporepressor, and that this is reversed in the presence of tryptophan, with the K_D of the WT now being lower than that of the mutant. However, more importantly, we discovered the existence of the stable population,[4] but were unwilling to speculate on its origin at that time. In the light of the present results, we now propose that this stable population may arise from the stabilization of the tandem complex by the mutation.

The flexibility of the reading heads is considerably dampened in the AV77 mutant, relative to the WT both in the presence, and in the absence of, the ligand.[19] Just as the ligand may act through its structuring of the tandem interface (in particular, helix E), the bracing of helix E by valine would also be expected to stabilize the tandem interface.

When we compare the complexes of apo~ and holorepressor; WT and AV77 TR; with each of the targets: *40-trpR*, *20-consensus*, and *40-EDCBA* (Figure 5) we observe that biexponential on-rates are always associated with biexponential off-rates, and vice versa.

Neither AV77 nor WT aporepressor show clear biexponential behavior with any of the DNA targets (Figure 5). Clearly, the ligand is required to extend the half-life of the complexes. WT holorepressor shows no biexponentiality with *40-trpR*, a small amount with *20-consensus*, and quite strongly with *40-EDCBA*. This parallels the suitability of these targets for multiple repressor binding. AV77 holorepressor exhibits biexponentiality for all three targets.

Despite the requirement for ligand in the stabilization of the half-life of the complex, it is important to note that the AV77 mutation acts to increase the apparent binding constant of the repressor to operator even in the absence of ligand.[4] That is, there are two distinct mechanisms by which the AV77 mutation becomes a super-repressor, which are kinetically distinguishable. One which alters the apparent K_D without appreciably lengthening the lifetime of the complex, and a second mechanism in which the lifetime of the complex is extended.

Fluorescence binding studies show that apo AV77 has increased binding to nonspecific DNA,[18] and therefore the earlier suggestion that the increased activity of AV77 is due to increased specificity for certain operators is unlikely to be correct.[20] Instead, we propose that apo AV77 has a conformation similar to that of the holoWT as suggested by Luisi and Sigler,[8] allowing it to conform to DNA generally more closely than does apo WT, but that it remains unable to form the tight, long-lived complex (Figure 6). Only when ligand is bound, do the reading heads adjust more closely to allow both tandem binding and the formation of sensitive, specific interactions with the bases.

Figure 5. SPR traces of WT and AV77, holo and aporepressor, binding to three different operators. Each pair of experiments (WT vs. AV77) had the same concentration of protein, and were (from top to bottom): 20 nM, 20 μM, 20 nM, 10 μM, 10 nM, 10 μM, respectively. The data are baseline corrected and normalized to the amount of DNA immobilized (such that each fully occupied operator site would theoretically generate 1 RU (normalized).

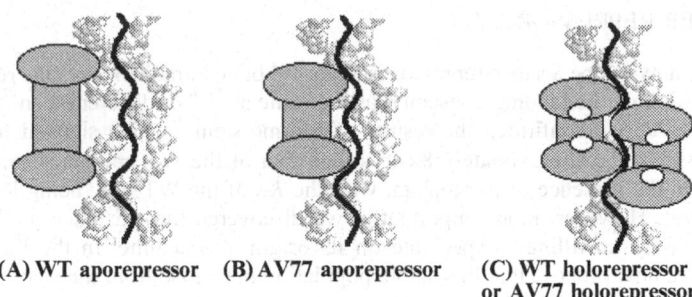

(A) WT aporepressor   (B) AV77 aporepressor   (C) WT holorepressor
                                                    or AV77 holorepressor

Figure 6. (A) The reading heads of WT aporepressor are unsuited to recognition of the major grooves of DNA, and so binding is poor. (B) AV77 aporepressor has the reading heads more suitably arranged for binding to DNA (nonspecifically), but is unable to form the correct contacts to bind in tandem to the DNA, and therefore the reading heads may not be exactly placed to form the contacts between the side chains and the base pairs which are required for specific binding to take place. (C) Holorepressor of both WT and AV77 are able to form tandem complexes. The reading heads mutually place each other in the precise orientation required to correctly contact the bases.

Based on fluorescence binding studies (an equilibrium method), Grillo and Royer[18] have proposed that AV77 acts as a super-repressor because its partially folded apo-repressor polymerizes to a lesser degree than apo-WT, leaving a larger population of functional dimers to bind productively to the operator DNA. Our kinetic measurements have shown however that at least two different mechanisms must be responsible for the super-repressor properties of AV77. First, there is greater binding of apoAV77 than of apoWT, which could be accounted for by the hypothesis of Royer and co-workers. However, we also observe a slower dissociation rate for holoAV77 – DNA complexes than for holoWT – DNA complexes. This cannot be accounted for by their hypothesis. We propose that the slowly dissociating species represents tetramers (and possibly higher oligomers) bound in tandem and that the greater tendency of the mutant holo-repressor to form functional oligomers bound in tandem makes a major contribution to its super-repressor properties.

We fit the holo AV77 – 40-EDCBA operator binding data to four alternative models (Figure 7). (1) A dimer (tandem) binding model, (2) a model in which two repressors bind to independent sites (parallel binding), and (3) A model in which a single dimer binds followed by a conformational change. At low protein concentrations, (0–12 nM), the dimer binding model (including mass transfer, but with no allowance for drift or a buffer refractive index contribution) gave reasonably good fits, yielding a χ^2 of 7.0. k_{a1} was ill-defined, as fixing it anywhere between 10^8 and 10^{10} $M^{-1}s^{-1}$ gave identical χ^2 values (k_{d1} and k_{a2} alter to compensate). Without allowing for mass transport, the χ^2 were 13.8, 9.7, and 9.9 for the tandem, two-site and conformational change models respectively (Figure 7). Although we are currently focusing on the data which indicate that the tandem interaction is required for long-term stability of the complex, it is also clear that the tandem interaction is not sufficient to explain all of the data. There are many factors which are also known, or are likely to play a role in repressor binding to DNA. These include (1) nonspecific binding at higher protein concentrations, (2) nonidentical binding sites which may have different affinities, and (3) a conformational rearrangement as the tandem complex forms, which may or not be rate-limiting. Further dissection of these contributions will require most likely the construction of

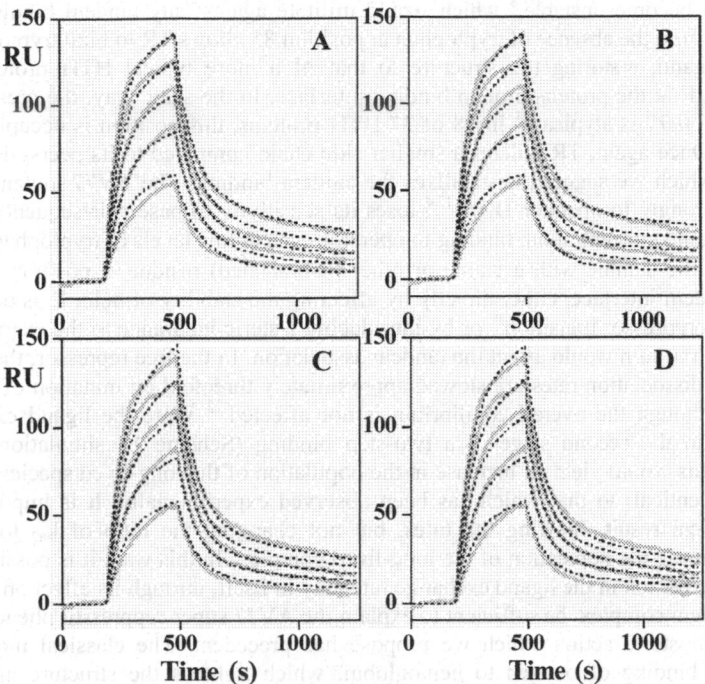

Time (s)   Time (s)

Figure 7. SPR traces of the binding of holo AV77 to; and the dissociation from, the *40-EDCBA* operator, at concentrations of 4, 6, 8, 10, and 12 nM (protein); with fits of the data to four different binding models. The *solid lines* are the experimental data, while the *dashed lines* are the fitted lines.

A: Dimers-masstransport, free fit ($k_{a1} = 2.0 \times 10^{10} \pm 2.6 \times 10^6$ M^{-1}s^{-1}, $kd_1 = 3.2 \times 10^2 \pm 4.1 \times 10^{-2}$ s^{-1}, $ka_2 = 1.4 \times 10^5 \pm 1.9 \times 10^1$ M^{-1}s^{-1}, $kd_2 = 6.5 \times 10^{-4} \pm 8.6 \times 10^{-8}$ s^{-1}, $R_{max} = 304 \pm 4.0 \times 10^{-2}$ RU, k_t (mass transport coefficient) = $1.5 \times 10^8 \pm 2.0 \times 10^4$. $\chi^2 = 6.97$).

B: Dimers, no mass transport ($k_{a1} = 6.1 \times 10^5 \pm 1.0 \times 10^{-2}$ M^{-1}s^{-1}, $k_{d1} = 1.0 \times 10^{-2} \pm 7.1 \times 10^{-5}$ s^{-1}, $k_{a2} = 1.2 \times 10^5 \pm 2.4 \times 10^3$ M^{-1}s^{-1}, $k_{d2} = 3.3 \times 10^{-4} \pm 3.1 \times 10^{-5}$ s^{-1}, $R_{max} = 300 \pm 3$ RU. $\chi^2 = 13.8$).

C: Parallel binding sites; ($k_{a1} = 2.9 \times 10^4 \pm 2.6 \times 10^3$ M^{-1}s^{-1}, $k_{d1} = 8.1 \times 10^{-4} \pm 2.3 \times 10^{-5}$ s^{-1}, $k_{a2} = 5.3 \times 10^5 \pm 9.7 \times 10^3$ M^{-1}s^{-1}, $k_{d2} = 1.4 \times 10^{-2} \pm 1.1 \times 10^{-4}$ s^{-1}, $R_{max1} = 520 \pm 49$ RU, $R_{max1} = 320 \pm 5$ RU. $\chi^2 = 9.72$).

D: Conformational change ($k_{a1} = 3.3 \times 10^5 \pm 5.5 \times 10^3$ M^{-1}s^{-1}, $k_{d1} = 1.3 \times 10^{-2} \pm 1.0 \times 10^{-4}$ s^{-1}, $k_{a2} = 1.3 \times 10^{-3} \pm 2.1 \times 10^{-5}$ M^{-1}s^{-1}, $k_{d2} = 1.1 \times 10^{-3} \pm 2.7 \times 10^{-5}$ s^{-1}, $R_{max} = 541 \pm 8$ RU. $\chi^2 = 9.91$).

new mutants and/or specifically designed operators to simplify the kinetics. We wish to emphasize that neither the parallel binding site nor the conformational change models are able to explain facts such as the inability of *40-trpR* operator DNA to stabilize the complex, in the manner seen for the *40-EDCBA* operator complexes, or the increased longevity of the SC5 complex with DNA. As far as we are aware, only tandem binding can explain these findings.

It is clear from the fact that several mutants can bind more effectively than the WT to DNA, that WT TR has not been selected during evolution for tightness of binding alone, but also for its ability to be regulated. The loss of a large side chain from position 85 (which is typically I, L, or W in HTH proteins[21]) is expected to cause the active conformation of the reading head to become unstable,[8] which would militate against any tandem binding. The cavity resulting from the absence of tryptophan at position 85 allows TR to bind tryptophan as a controlling ligand, restoring the structure to that of a more typical HTH protein, and stabilizing helix E at the protein–protein binding interface. In the same way, the presence of alanine at position 77 is atypical.[21] In 18 of 37 HTH proteins, this position is occupied by a valine residue. Once again, TR utilizes a smaller side chain compared to its peers, destabilizing helix E[19] which we suggest, destabilizes the tandem binding. The AV77 mutant, while binding more strongly to operator DNA,[17,4] loses its sensitivity to base pair sequence[18] and, separately, to ligand concentration, binding to operator even at low levels of tryptophan.[14]

Substituting the alanine with a valine or other (β-branched) residue at position 77 may stabilize the tandem interface, either directly by affecting the stability of helix E as occurs in the WT apo/holorepressor transition[19] or by introducing a steric hindrance to the exchange of the ligand, which in turn would affect the tandem association. In the free repressor, the ligand association and dissociation rates are slowed approximately threefold on mutation of alanine 77 to valine, although the overall equilibrium is not affected.[22] Were the ligand exchange directly linked to the second stage of a two-step binding (Scheme 1), simulations show (Figure 8) that this would yield an increase in the population of the long-lived species similar (although not identical) to that which has been observed experimentally. It is important to restate this critical result. Slowing the rates, but not changing the ratio of k_{a2} to k_{d2}, is sufficient to increase the population of the long-lived complex. In this way; it is possible that the observed[22] reduction in the ligand exchange rate may in itself, through its effect on the stability of the tandem complex, be sufficient to explain the AV77 super-reppressor phenoltype.

The mode of allosteric action which we propose has precedent. The classical model for allostery is the binding of oxygen to hemoglobin, which tightens the structure and consequently reorganizes the intersubunit interface (the inter-β-chain distance decreases by 7 Å on oxygenation). It is possible that this proposed modulation of oligomerization by ligand binding also occurs in other HTH–DNA-binding proteins. Inspection of the structure of the met-repressor (PDB 1CMA) indicates that the same combination of elements are present (inter-dimer interfacial helices – residues 54–66 in this case – which form part of the ligand-binding pocket), and that therefore the activation of met-repressor by S-adenosyl methionine may occur in a similar fashion. It would be of interest to see if the dynamics of these helices are altered on ligand binding, as is helix E of TR.

In conclusion, we have observed that a mutant which is designed to enhance the tandem interaction of TR on DNA results in the most stable complex demonstrated to date. This confirms that it is possible for the similar longevity of the AV77–DNA complex to also be due in part to an increased stability of the tandem complex (although we do not claim to have proven this). We have also suggested a novel role for the ligand in the activation of TR, based on our results, which is consistent with and may help explain a wide variety of experimental findings in the literature.

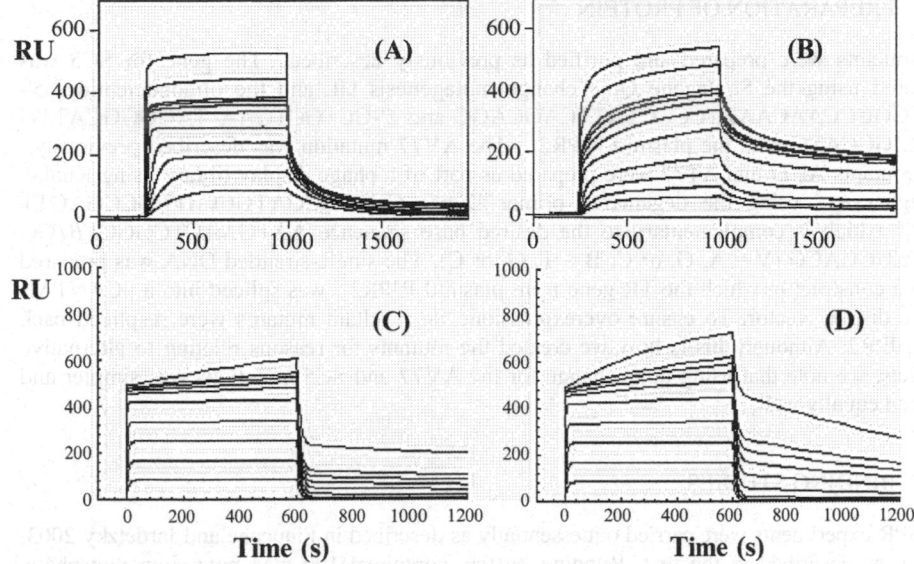

Figure 8. Experimental and simulated SPR data. Upper figures; experimental data (taken from Finucane and Jardetzky[4]) **(A)** holo-WT–20-consensus interaction, **(B)** holo AV77–20-consensus interaction. Lower figures; simulation of the same protein concentrations binding to DNA, using the dimer binding model, with $k_{a1} = 10^7$; $k_{d1} = 0.1$, and **(C)** $k_{a2} = 10^3$, $k_{d2} = 3 \times 10^{-4}$; **(D)** $k_{a2} = 3 \times 10^3$, $k_{d2} = 9 \times 10^{-4}$. In **(C)** and **(D)**, the equilibrium ratio between k_{a2} and k_{d2} has been unaffected. Although the simulation does not perfectly fit the data, the figure demonstrates how a reduction in the ligand exchange rate may result in stabilization of the TR-operator DNA complex.

3. Materials and Methods

3.1. MATERIALS

Oligonucleotides were obtained (highly purified salt-free) from MWG Biotech (High Point, NC) and used without further purification. Double-stranded operators were annealed by mixing both strands at a concentration of 10 µM, before heating at 95° for 5 minutes, followed by cooling to 20°C at 1°/min. The concentration was then reduced to 4 µM for storage. The operators were further diluted to 200 nM just prior to immobilization on the Biacore SA chips. Restriction enzymes were obtained from New England Biolabs (Cambridge, MA). Quickchange mutagenesis kits were obtained from Stratagene. SPR chips were obtained from BIAcore.

3.2.   PREPARATION OF PROTEIN

All mutants were prepared and purified as previously described.[4] The gene for SC5 was prepared using the Stratagene Quickchange mutagenesis kit, and the oligonucleotides 5'-CATGGCCCAACAA*TGC*CCCTATTCAGCAGC and 5'-GCTGCTGAA TAGGGGCATTG TTGGGCCATG with the plasmid PJPR2.[23] The AV77 mutation was described previously.[4] The mutants AL77 and AT77 were prepared as part of a phage display library using Kunkel mutagenesis,[24] using the degenerate primer Dyn5: CGTCGCGATGCC*AVB*GCCGAGTT CATT which is complementary to the desired base sequence AATGAACTCGGC*VBT*GG CATCGCGACG (V = A, G, or C; B = T, G, or C). The single-stranded DNA was prepared from a construct in which the TR gene from plasmid PJPR2[23] was spliced into a pCANTAB phage display vector. To ensure overexpression, the resultant mutants were respliced back into pJPR2. Although this is how we created the mutants for reasons relating to alternative projects, we note that using quickchange for the AV77 and SC5 mutations was simpler and worked equally well.

3.3.   BINDING STUDIES

All SPR experiments were carried out essentially as described in Finucane and Jardetzky 2003, except as described in the text. Running buffers contained 100 mM potassium phosphate, 20 mM KCl, and 0.005% Tween 20 at pH 7.5, with the addition of 2 mM tryptophan and/or 10 mM DTT as indicated in the text. All experiments were run at 25°C, with a flow rate of 50 µL/min. The level of DNA immobilized in all cases was less than 200 RU to reduce mass transport effects (although they were not eliminated). As before, the data were zeroed by subtracting data taken simultaneously from a lane in which no DNA had been bound to the surface. The data were fit to two models included (two-state reaction{conformation change} and heterogeneous ligand, parallel reaction) in the BIAevaluation software provided with the instrument (BIAeval 3.01), and to a third developed by us to fit the dimeric binding model (Scheme 1).

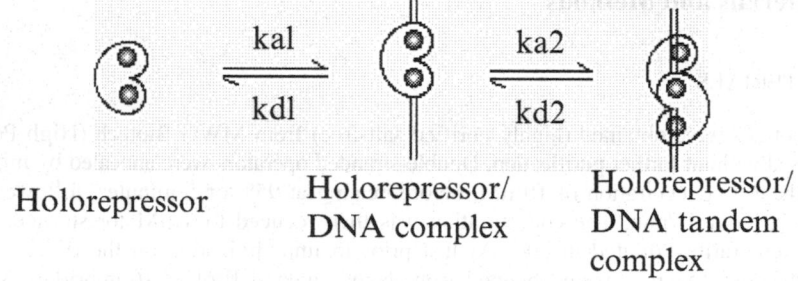

3.4.   SIMULATIONS

SPR data were simulated using the program Mathematica V 2.2.2 on Macintosh; and the following differential equations (see also Scheme 1); (the code is available from the authors on request).

o'[t] == −freeprot ka1 o[t] + kd1 ho[t],
ho'[t] == kd2 hho[t] + ka1 freeprot o[t] − kd1 ho[t] − ka2 ho[t],
hho'[t] == ka2 ho[t] − kd2 hho[t],
hho[0] == 0,
ho[0] == 0,
o[0] == Rmax

where o = operator DNA; ho = holorepressor-operator; and hho is the dimeric complex; free-prot is the free protein concentration and is constant as it is constantly being renewed by flushing the cell. (Derivatives are denoted by prime (') symbols.)

4. Acknowledgments

This work was funded under grant 5RO1 GM33385-16 from the NIH.

References

1. Kumamoto, A.A., Miller, W.G., and Gunsalus, R.P. (1987). *E. Coli* tryptophan repressor binds multiple sites within the *aroH* and *trp* operators. *Genes Dev.* 1, 556–564.
2. Lawson, C.L. and Carey, J. (1993). Tandem binding in crystals of a *trp* repressor/operator half-sie complex. *Nature* 266, 178–182.
3. Liu, Y-C. and Matthews, K.S. (1993). Dependence of *trp* repressor-operator affinity, stoichio-metry, and apparent cooperativity on DNA sequence and size. *J. Biol. Chem.* 268, 23239–23249.
4. Finucane, M.D. and Jardetzky, O. (2003). Surface plasmon resonance studies of wild-type and AV77 tryptophan repressor resolve ambiguities in super-repressor activity. *Protein Sci.* 12, 1613–1620.
5. Otwinowski, Z., Schevitz, R.W., Zhang, R.G., Lawson, C.L., Joachimiak, A., Marmorstein, R.Q., Luisi, B.F., and Sigler, P.B. (1988). Crystal structure of *trp* repressor/operator complex at atomic resolution. *Nature* 335, 321–329.
6. Zhang, H., Zhao, D., Revington, M., Lee, W., Jia, X., Arrowsmith, C., and Jardetzky, O. (1994). The solution structures of the trp repressor-operator DNA complex. *J. Mol. Biol.* 238, 592–614.
7. Ivanetich, K.M. and Santi, D.V. (1992). 5,6-dihydropyrimidine adducts in the reactions and interactions of pyrimidines with proteins. *Prog. Nucleic Acid Res. Mol. Biol.* 42, 127–156.
8   Luisi, B.F. and Sigler, P.B. (1990). The stereochemistry and biochemistry of the trp repressor-operator complex. *Biochim. Biophys. Acta* 1048, 113–126.
9. Zhao, D., Arrowsmith, C.H., Jia, X., and Jardetzky, O. (1993). Refined solution structures of the *Escherichia coli trp* holo- and aporepressor. *J. Mol. Biol.* 229, 735–746.
10. Finucane, M.D. and Jardetzky, O. (1995). Mechanism of hydrogen-deuterium exchange in *trp* repressor studied by 1H-15N NMR. *J. Mol. Biol.* 253, 576–589.
11. Howard, A.E. and Kollman, P.A. (1992). Molecular dynamics studies of a DNA-binding protein: 1. A comparison of the *trp* repressor and *trp* aporepressor aqueous simulations. *Protein Sci.* 1, 1173–1184.
12. Moseley, H.N., Lee, W., Arrowsmith, C.H., and Krishna, N.R. (1997). Quantitative determination of conformational, dynamic, and kinetic parameters of a ligand-protein/DNA complex from a complete relaxation and conformational exchange matrix analysis of intermolecular transferred NOESY. *Biochemistry* 36, 5293–5299.

13. Kelley, R.L. and Yanofsky, C. (1985). Mutational studies with the *trp* repressor of *Escherichia coli* support the helix-turn-helix model of repressor recognition of operator DNA. *Proc. Natl. Acad. Sci. USA* 82, 483–487.

14. Arvidson, D.N., Pfau, J., Hatt, J.K., Shapiro, M., Pecoraro, F.S., and Youderian, P. (1993). Tryptophan super-repressors with alanine 77 changes. *J. Biol. Chem.* 268, 4362–4369.

15. Hurlburt, B.H. and Yanofsky, C. (1990). Enhanced operator binding by *trp* superrepressors of *Escherichia Coli*. *J. Biol. Chem.* 265, 7853–7858.

16. Liu, Y-C. and Matthews, K.S. (1994). *trp* Repressor mutations alter DNA complex stoichiometry. *J. Biol. Chem.* 269, 1692–1698.

17. Marmorstein, R.Q., Sprinzl, M., and Sigler, P.B. (1991). An alkaline phosphatase protection assay to investigate *trp* repressor/operator interactions. *Biochemistry* 30, 1141–1148.

18. Grillo, A.O. and Royer, C.A. (2000). The basis for the super-repressor phenotypes of the AV77 and EK18 mutants of *trp* repressor. *J. Mol. Biol.* 295, 17–28.

19. Gryk, M.R. and Jardetzky, O. (1996). AV77 hinge mutation stabilizes the helix-turn-helix domain of *trp* repressor. *J. Mol. Biol.* 255, 204–214.

20. Gryk, M.R., Jardetzky, O., Klig, L.S., Yanofsky, C. (1996). Flexibility of DNA binding domain of trp repressor required for recognition of different operator sequences. *Protein Sci.* 5, 1195–1197.

21. Brennan, R.G. and Matthews, B.W. (1989). The helix-turn-helix DNA binding motif. *J. Biol. Chem.* 264, 1903–1906.

22. Schmitt, T.H., Zheng, Z., and Jardetzky, O. (1995). Dynamics of tryptophan binding to *Escherichia coli* trp repressor wild type and AV77 mutant: An NMR study. *Biochemistry* 34, 13183–13189.

23. Paluh, J.L. and Yanofsky, C. (1986). High-level production and rapid purification of the *E. coli trp* repressor. *Nucl. Acids. Res.* 14, 7851–7860.

24. Kunkel, T.A., Bebenek, K., and McClary, J. (1991). Efficient site-directed mutagenesis using uracil-containing DNA. *Methods Enzymol.* 204, 125–139.

BASIC PRINCIPLES OF RNA NMR SPECTROSCOPY

PETER J. LUKAVSKY
Laboratory of Molecular Biology
Medical Research Council, Hills Road,
Cambridge, CB2 2QH, UK
pjl@mrc-lmb.cam.ac.uk
Phone: 00-44-1223-402417
Fax: 00-44-1223-213556

Abstract: Structure determination of RNA molecules at the atomic level is necessary to rationalize their multiple cellular functions. NMR spectroscopy is a powerful method to determine RNA solution structures up to 30kDa. Here I give an overview of the standard NMR techniques required for high-resolution structure determination of RNAs up to 15kDa. In addition, I will discuss which problems are encountered when studying larger RNAs by NMR (up to 100kDa) and present a generally applicable approach for the high-resolution NMR structure determination of RNAs up to 30kDa.

1.   Introduction

RNA plays a central role in many important biological processes ranging from protein synthesis to RNA processing. Consisting of only four individual building blocks, RNA is nevertheless an extremely versatile biomolecule, able to interact with a large variety of small ligands as well as macromolecules in diverse, yet highly specific ways. Structure determination of biological RNAs and their complexes with proteins and other ligands is necessary to reveal the structural basis of the multiple biological functions of RNA. The first two high-resolution structures of stable RNA hairpins (Cheong et al., 1990; Heus and Pardi, 1991b) determined by nuclear magnetic resonance (NMR) spectroscopy in the early 1990s mark the beginning of a fruitful decade for RNA NMR. During this period numerous NMR techniques were developed to study RNA structure, but during the late 1990s, further progress stalled at a molecular mass limit of 15kDa (Varani et al., 1996). Novel isotopic labeling schemes (Kim et al., 2002; Scott et al., 2000) and the use of angular restraints derived from residual dipolar couplings (RDCs) (Hansen et al., 1998; Tjandra and Bax, 1997) made it possible to cross this "15kDa-barrier" very recently and nowadays NMR structure determination of RNAs up to 30kDa is feasible, but still challenging (Latham et al., 2005; Lukavsky and Puglisi, 2005; Tzakos et al., 2006b). Here, I describe the general aspects of RNA NMR spectroscopy in contrast to protein NMR and discuss specific RNA NMR experiments for structure determination of RNAs up to 15kDa. In addition, a generic protocol for the structure determination of larger RNAs up to 30kDa and its first successful application to the 25kDa hepatitis C viral (HCV) internal ribosome entry site (IRES) domain II RNA will be presented together with a brief outlook into future developments.

J. D. Puglisi (ed.), Structure and Biophysics – New Technologies for Current Challenges in Biology and Beyond, 65–80.

2.   General Aspects of RNA Structure Determination by NMR Spectroscopy

High-resolution RNA or protein solution structure determination by NMR spectroscopy follows similar procedures. In both cases, near complete resonance assignment is required and this necessitates uniform isotopic labeling of the biomolecules with ^{13}C and ^{15}N. Double and triple resonance experiments are used to identify individual spin systems, classify them by residue type and subsequently connect neighboring residues to obtain sequential assignments of the biomolecule. The structural information is then obtained from nuclear Overhauser effect spectroscopy (NOESY)-type experiments that give distance information between adjacent protons (<6 Å) and NMR experiments that measure torsion angles for the protein or RNA backbone. The maximum achievable number of local restraints is then used in a simulated annealing protocol followed by restrained molecular dynamics (MD) calculations to generate structural ensembles, which satisfy the distance and torsion angle restraints.

Despite this general similarity, RNA NMR spectroscopy is quite distinct from protein NMR spectroscopy. RNA is composed of only 4 different nucleosides, namely guanosine, cytidine, adenosine, and uridine, with an average molecular weight of 340 Da. In contrast, proteins are composed of 20 different amino acids with an average molecular weight of 130 Daltons (Da). The molecular weight difference of the RNA or protein building blocks results in a lower proton density for RNA and therefore less structural restraints per residue can be extracted compared to proteins. Near complete resonance assignment is therefore pivotal for RNA in order to obtain the maximum number of distance restraints possible. Another consequence of the smaller number of RNA building blocks is a less favorable chemical shift dispersion for all RNA nuclei and this makes unambiguous resonance assignments more difficult in RNA. Proteins, on the other hand, often display a backbone amide nitrogen chemical shift dispersion of about 30 ppm. Most protein triple resonance 3D and 4D NMR experiments used for sequential resonance assignments therefore take advantage of this favorable chemical shift dispersion by correlating less well-dispersed amino acid side-chain resonances to the well-dispersed amide nitrogen in the third dimension.

The RNA backbone, by contrast, does not contain a nucleus with such favorable chemical shift dispersion, since both carbon (^{13}C) and phosphorus (^{31}P) nuclei along the RNA backbone resonate within narrow chemical shift windows (Table 1). In addition, the unfavorable chemical shift anisotropy parameters of the ^{31}P nuclei also result in short transverse relaxation times hampering efficient magnetization transfer along the backbone in RNA triple resonance experiments in comparison with proteins. Furthermore, magnetization transfer along the protein backbone in triple-resonance experiments is more efficient, since it utilizes much larger coupling constants in the range of 10 to 55 Hz compared to RNA, whose backbone two- and three-bond couplings range from 3 to 10 Hz. Sequential assignments in RNA are therefore almost always supplemented by assignments obtained from NOESY-type experiments.

However, not only the backbone of RNA and proteins have very different NMR spectroscopic properties; both biomolecules also differ in their side-chains. Proteins contain both aromatic and aliphatic side-chains, ranging from acidic to basic and hydrophilic to hydrophobic, while RNA side-chains are composed of only four different aromatic bases, which differ from the 20 different protein side-chains. Therefore RNA base-specific NMR experiments for the unambiguous identification of their resonances are required. These experiments, which correlate protons and/or heteronuclei within the aromatic ring system of the individual bases, seldom yield a complete set of assignments, since correlations are often

TABLE 1. Chemical shift ranges for spin-1/2 nuclei in RNA

^{1}H chemical shift ranges (ppm)	Proton type
3.5 – 5	ribose H-2', H-3', H-4', H-5', H-5"
5 – 6.2	ribose H-1', pyrimidine H-5
6.5 – 7.5	ribose 2'-OH
6 – 9	adenine, guanine, cytosine amino
6.5 – 8.5	aromatic purine H-8, H-2 and pyrimidine H-6
9.5 – 15	guanine and uracil imino

^{13}C chemical shift ranges (ppm)	Carbon type
60 – 70	ribose C-5'
70 – 80	ribose C-2' and C3'
80 – 85	ribose C4'
85 – 95	ribose C1'
95 – 100	cytosine C-5
100 – 105	uracil C-5
115 – 120	purine C-5
135 – 145	purine C-8 and pyrimidine C-6
147 – 150	purine C-4
150 – 157	adenine C-2
155 – 159	purine C-6
165 – 170	pyrimidine C-4

^{15}N chemical shift ranges (ppm)	Nitrogen type
65 – 70	guanine amino N-2
75 – 80	adenine amino N-6
90 – 95	cytosine amino N-4
140 – 145	uracil anomeric N-1
145 – 150	guanine imino N-1, cytosine anomeric N-1
155 – 165	uracil imino N-3
160 – 170	guanine and adenine anomeric N-9
190 – 195	cytosine N-3
210 – 215	adenine N-3
215 – 225	adenine N-1
225 – 235	purine N-7

^{31}P chemical shift ranges (ppm)	Posphorus type
(-2) – (-4)	phosphodiester backbone

lost due to inefficient magnetization transfer through the small heteronuclear couplings within the aromatic bases. In addition, imino and amino protons of the uracil and guanine bases experience exchange processes that often make their detection difficult. Nonhydrogen bonded imino protons usually exchange rapidly (millisecond timescale) with solvent, whereas amino protons experience additional chemical exchange from rotation about C–N bond, in particular for guanine and adenine amino groups. Detection of exchangeable protons in RNA therefore requires not only water suppression schemes that limit the amount of solvent saturation and thus signal loss from saturation, but also NMR techniques which allow efficient magnetization transfer for amino protons experiencing intermediate chemical exchanges, like Carr–Purcell–Meiboom–Gill (CPMG) pulse trains and heteronuclear TOCSY mixing schemes.

Following backbone and side-chain assignments, structural information is derived from NOESY-type experiments, which provide distance restraints and from 2D and 3D double and triple resonance NMR experiments, which yield qualitative torsion angle information. The precision of the structural ensemble depends both on the quantity and "quality" of the restraints. For example, long-range NOEs, between residues, which are far apart in sequence defining secondary structure elements (e.g. base pairs) or tertiary interactions (e.g. helix–helix packing), add more to the global structural precision than intraresidual or sequential NOEs (Allain and Varani, 1997). In general, global precision is more easily achieved for proteins than for RNA, since RNA molecules often adopt extended conformations and therefore yield only a limited number of long-range restraints in comparison with globular, compactly folded proteins. Moreover, the naturally lower proton density of RNA molecules aggravates this problem, since this additionally reduces the number of obtainable restraints per residue. The application of RDCs to biomolecular NMR (Bax et al., 2001), greatly improved the global precision of NMR structures and especially benefited both the local and global precision of RNA structural ensembles determined by NMR spectroscopy (Hansen et al., 1999; Mollova and Pardi, 2000; Warren and Moore, 2001). RDC-derived angular restraints also allow to define the global conformation of larger RNAs, and their recent application to the structure determination of three large RNAs (Davis et al., 2005; D'Souza and Summers, 2004; Lukavsky et al., 2003) has assisted to push the molecular mass limit of RNA NMR beyond 30kDa.

To summarize, state-of-the-art high-resolution RNA structure determination by NMR spectroscopy requires not only conventional distance and torsion angle restraints, but also angular restraints derived from RDCs to define both local and global RNA conformation with high precision. In order to extract a maximum number of restraints, fully RNA-optimized NMR experiments are required that consider the less favorable NMR properties of RNA oligonucleotides compared to proteins. The ideal RNA NMR experiments, which I will describe in the following section, use very basic and therefore short pulse sequences to minimize the loss of signal due to short transverse relaxation times while still providing maximum resolution for the poorly dispersed RNA nuclei to allow unambiguous resonance assignment; the basis for RNA structure determination by NMR.

3.   RNA NMR Experiments

A high-resolution RNA structure determination requires almost complete ^{1}H, ^{13}C, ^{15}N, and ^{31}P resonance assignments. Individual spin systems of the ribose and base moieties of each residue need to be assigned unambiguously and linked together using triple resonance experiments, which link ribose and base within one nucleoside and sequential nucleosides through the intervening phosphate group. Unambiguous assignment of all resonances belonging to one residue is achieved by transferring magnetization between covalently bonded nuclei. Through-bond NMR methods are therefore used to assign both ribose spin systems and base resonances. Resonance overlap often requires supplementing this information with additional assignments derived from NOESY-type experiments, which also yield proton–proton distance information for structure calculations. Some through-bond NMR experiments also yield quailtative torsion angle information, which can be derived from heteronuclear and homonuclear J coupling dependent cross peak intensities. This information is also used in structure calculations. Finally, one-bond C–H or N–H J couplings are measured in aqueous solution and liquid crystalline medium (typically by adding 10mg/ml Pf1 phage). The latter induces weak alignment of the RNA relative to the magnetic field and this gives rise to residual dipolar couplings, which provide valuable angular restraints for the refinement of structural ensembles calculated with distance and torsion angle restraints alone.

3.1. RNA NMR EXPERIMENTS FOR THE ASSIGNMENT OF SPIN SYSTEMS

Ribose and pyrimidine base spin system assignments use 2D TOCSY, 2D DQF-COSY, and 3D HCCH-type experiments, which can be performed both with D_2O samples (100% D_2O) or water samples (95% H_2O/5% D_2O) provided efficient water suppression techniques are applied. Nevertheless, some water suppression techniques, such as presaturation of the H2O signal, can also lead to loss of RNA resonances around the water signal. Therefore, in most applications, D_2O samples are the preferred option, since no water suppression of the residual HOD resonance is required. For the homonuclear 2D TOCSY and DQF-COSY experiments, the WET water suppression scheme (Smallcombe et al., 1995) works best, while 3D HCCH-type experiments employ gradient-based water suppression. Both homonuclear experiments yield cross peaks for each pyrimidine spin system (H5 and H6 protons) and sugar pucker information from ribose H-1' –H-2' scalar couplings and cross peak intensities. In case of favorable ribose H-5' and H-5" chemical shift dispersion, γ torsion angle information can be extracted from 2D DQF-COSY experiments, if ^{31}P-decoupling is used both during t_1-periods and the acquisition time. Heteronuclear 3D HCCH-type experiments, namely 3D HCCH-TOCSY (Kay et al., 1993), 3D HCCH-COSY (Kay et al., 1990), and 3D HCCH-RELAY (Pardi and Nikonowicz, 1992) yield ^{13}C-resolved spin system assignments. Both the 3D HCCH-TOCSY and COSY experiments correlate pyrimidine or ribose spin systems, since these spin systems require different ^{13}C frequency offsets (see Table 1) and transfer delay times. The 3D HCCH-RELAY experiment adds an extra ^{13}C–^{13}C transfer to the HCCH-COSY and thereby allows magnetization transfer through two carbon–carbon bonds (Kay et al., 1990). The 3D HCCH-TOCSY is also commonly used to correlate intraresidue adenine H2 and H8 resonances (Marino et al., 1994a). This correlation requires transfer times of up to 100 ms due to the small ^{13}C–^{13}C two-bond carbon couplings (10 Hz) within the adenine base moiety. This experiment often fails with RNA molecules beyond 30 nucleotides (nt) due to their slower overall tumbling times, which lead to shorter transverse relaxation times and loss of signal during the 100ms TOCSY mixing. For larger RNAs, a transverse relaxation optimized (TROSY) relayed HCCH-COSY is therefore often the better choice for identifying the H2 and H8 protons belonging to one residue (see below) (Simon et al., 2001).

Intranucleotide correlations between exchangeable and nonexchangeable protons of the base moiety such as 2D cytosine and uracil H(NCCC)H experiment (Simorre et al., 1995), 2D adenine (Simorre et al., 1996b), and guanine (Simorre et al., 1996a) hetero-TOCSY experiments, are used to identify all protons belonging to one base residue. All these correlation experiments use multiple heteronuclear transfer steps (up to 100 ms) including at least one small one-bond (8–10Hz) ^{15}N–^{13}C or ^{13}C–^{13}C coupling and therefore often fail for RNAs larger than 40 nucleotides, when transverse relaxation times are too short. For larger RNAs, the 3D TROSY relayed HCCH-COSY experiment can be used to correlate adenine H2/H8 resonances (Simon et al., 2001). The pathway of the out-and-back-type magnetization transfer is explained in Figure 1A. After initial magnetization transfer through the one-bond H2/8–C2/8 couplings, magnetization is further transferred through the small C2–C5 or C8–C4/6 ^{13}C–^{13}C couplings, and then relayed to C4/C6 or C5, respectively. The 3D experiment contains one-proton plane (aromatic H2/H8) and two-carbon planes corresponding to the directly attached carbon (C2/C8) and the relayed HCCH-type correlations, respectively (Figure 1C and D). In the 2D ^{13}C–HSQC of a 40nt RNA shown in Figure 1B, a C2–H2 and a C8–H8 cross peak belonging to the same residue are labeled and the corresponding correlation planes are shown in Figure 1C and D. In the adenine C2-plane (Figure 1C), two correlations to C4 and C6 arise from the first ^{13}C–^{13}C transfer, and one cross peak from the relayed transfer to C5, while in the adenine C8-plane (Figure 1D), one correlation to C5 arises from the first ^{13}C–^{13}C transfer, and two cross peak from the relayed transfer to C4 and C6. Matching carbon chemical shifts of the C4/C5/C6 correlation peaks of both the H2 and H8 resonance unambiguously assigns the latter to the same adenine residue. Since the final C–H transfer employs a TROSY-type transfer, this experiment is remarkably robust and worked, in our hands, also for a 67nt RNA.

Linkage between the base and sugar moiety is achieved using triple resonance multiple-quantum or TROSY-type HCN experiments (Fiala et al., 2000; Marino et al., 1997) that

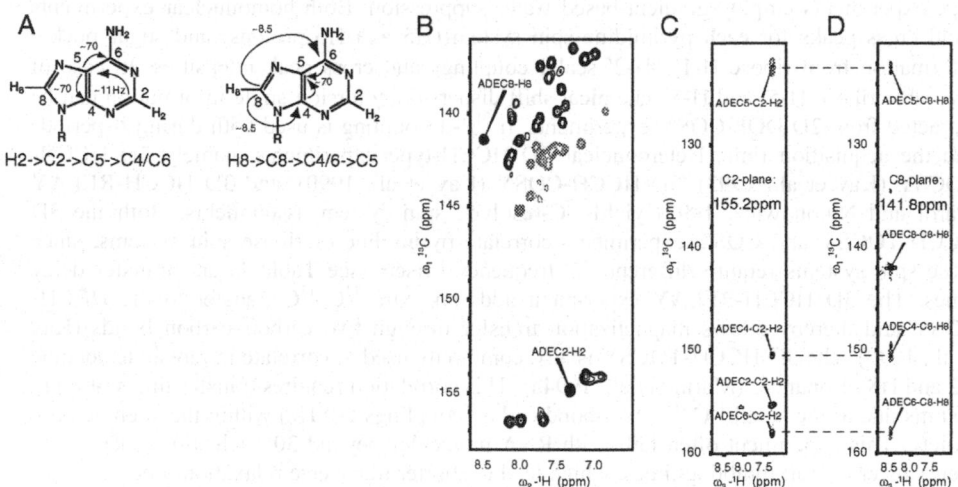

Figure 1. Adenine H2–H8 correlation using the 3D TROSY relayed HCCH-COSY experiment. (**A**) Magnetization transfer during the TROSY relayed HCCH-COSY experiment. Relevant couplings are indicated. (**B**) Aromatic region of a 2D ^{13}C CT-HSQC of a 40nt RNA. The resonances belonging to one residue are indicated. (**C**) and (**D**) C2-plane and C8-plane of the 3D TROSY relayed HCCH-COSY spectrum recorded on the same RNA as in (**B**). Correlations to C4/C5/C6 are labeled and dashed lines indicate matching chemical shifts that unambiguously correlate the adenine H2–H8 resonances.

connect the aromatic base proton to the H1' ribose proton through the anomeric nitrogen, N9 for pyrimidines and N1 for pyrimidines, respectively. The experiment can be performed as a 2D or 3D experiment; the later yields additional resolution with ^{13}C in the third dimension. Multiple variations of these experiments, some of them with much improved sensitivity, exist but all of them suffer from the same limitation, the poor ^{15}N chemical shift dispersion of the anomeric N9/N1 resonances, which makes it difficult to extract unambiguous assignments without additional information from NOESY-type experiments.

3.2. RNA NMR EXPERIMENTS FOR DETERMINATION OF BACKBONE TORSION ANGLES

Individual nucleosides are linked by a combination of two through-bond experiments involving the ^{31}P nucleus, namely the ^1H, ^{31}P COSY(Sklenar et al., 1986) and the 3D HCP experiment (Marino et al., 1994a, b), both of which also yield qualitative β and ε torsion angle information. Both experiments are best performed on a 100% D_2O sample, since any type of water suppression scheme that efficiently suppresses the water signal, also attenuates ribose protons and thereby affects the sensitivity of the experiments, especially of the 3D HCP experiment. The ^1H, ^{31}P COSY experiment links individual nucleosides through H-3'(i-1)–P(i) correlations. Additional intranucleoside H-4'(i)–P(i) and H-5/5'(i)–P(i) correlations can be obtained in flexible backbone regions and for gauche$^+$ conformation of torsion angle β. The ^1H, ^{31}P COSY spectrum of HCV IRES domain IIId, a 29nt RNA hairpin (Lukavsky et al., 2000), in Figure 2 shows a strong sequential H-3'(i-1)–P(i) and weak H-5'(i)–P(i) correlations for A257 indicating a trans conformation of torsion angle β. G258, on the other hand is in a gauche$^+$ conformation around torsion angle β, which yields a strong H-5''(i)–P(i) correlation and a much weaker sequential cross peak. Sequential assignments are also obtained from the 3D HCP experiment, which yields β and ε torsion angle dependent cross peak intensities for H-2'(i-1)–C-2'(i-1)–P(i), H-3'(i-1)–C-3'(i-1)–P(i), H-4'(i-1)–C-4'(i-1)–P(i), H-4'(i)–C-4'(i)–P(i) and H-5'/5'(i)–C-4'(i)–P(i) correlations (Marino et al., 1999). The interpretation of 3D HCP data is straightforward as shown in the following example: when torsion angle ε adopts a trans conformation, as encountered in regular A-form RNA, two cross peaks of equal intensity corresponding to the intraresidue (i) and inter-residue (i-1) correlation between ribose C4' and the linking phosphate are observed (Figure 4A). In Figure 4B, the correlations characteristic

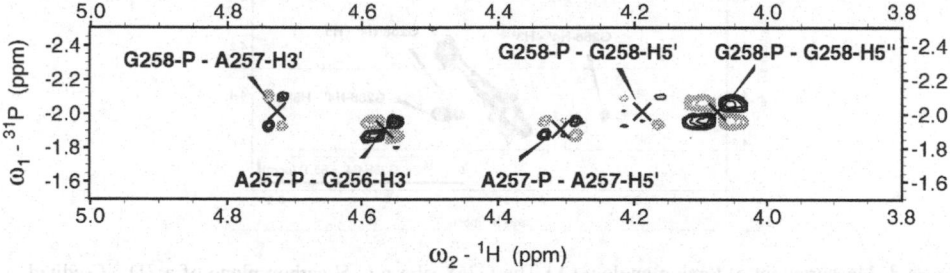

Figure 2. Measurement of torsion angle β (**A**) Portion of a $^1H,^{31}P$ COSY spectrum collected at 25°C showing torsion angle-dependent $^1H,^{31}P$ cross peaks for A257 and G258.

of a gauche⁻ conformation of torsion angle ε are shown. Instead of the strong inter-residue (i-1) correlation between ribose C4′ and the linking phosphate, the inter-residue (i-1) C2′ and C3′ correlations are observed in addition to the intraresidue (i) C4′–P cross peak. It should be noted that we always use the first published, very basic 3D HCP experiment, which only works with D_2O samples, because it lacks any water suppression scheme (Marino et al., 1994b). Newer versions of this experiment, which also includes sophisticated water suppression and sensitivity enhancement schemes, yielded about 1/5 of the signal to noise compared to the very basic version (using the same D_2O sample!!), since the additional INEPT transfer of the sensitivity enhancement pulse train lengthens the experiment and leads to significant signal loss due to the notoriously short transverse relaxation times of RNAs larger than 30nt. The last RNA backbone experiment is the 3D HMQC-TOCSY, which yields ^{13}C ribose carbon resolved proton–proton TOCSY correlations, whose intensity yields qualitative sugar pucker and γ torsion angle information. In regular A-form RNA, where backbone angle γ is in gauche⁺ conformation, no proton–proton correlation is observed between H-4′ and H-5′/H5″ protons due to the small three bond coupling constant of 1–3 Hz. For the loop residue G268 of HCV IRES domain IIId (Lukavsky et al., 2000), in contrast, H-4′ to H-5′ and H-5″ correlations could be observed in both the C-5′ and the C-4′ plane (Figure 3). Both carbon planes also showed a stronger correlation to the H-5″ proton indicative of a trans conformation of this γ torsion angle.

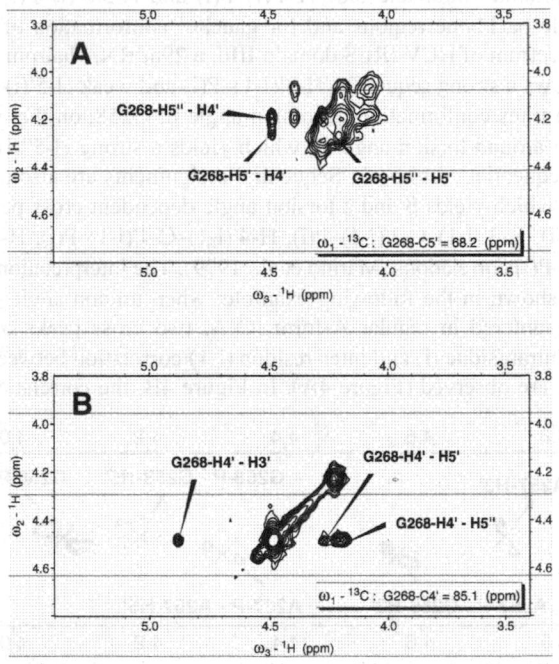

Figure 3. Measurement of torsion angle γ (**A**) The G268 ribose C-5′ carbon plane of a 3D ^{13}C-edited HMQC TOCSY spectrum collected at 25°C showing ribose H-4′ to H-5′/H-5″ correlations indicative of a trans γ torsion angle information (see text). (**B**) The G268 ribose C-4′ carbon plane of the same spectrum showing the same correlations as in (A).

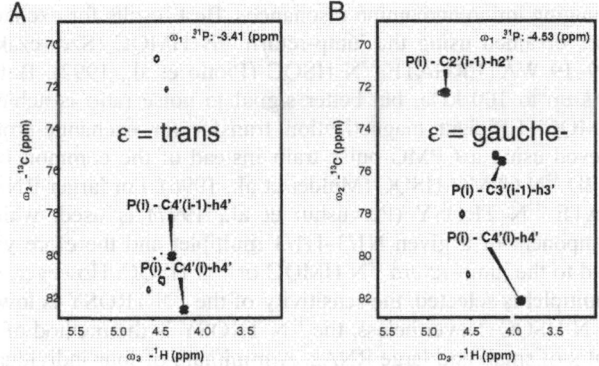

Figure 4. Measurement of torsion angle ε **(A)** Phosphorus plane of a 3D HCP spectrum showing two correlations between the intra-residue and inter-residue ribose C4' and the intervening phosphate indicative of a trans ε torsion angle (see text). **(B)** Phosphorus plane of a 3D HCP spectrum showing one intra-residue ribose C4'–P correlation and two sequential correlations of the same phosphate to ribose C2'/C3' indicative of a gauche⁻ ε torsion angle (see text).

3.3. RNA NMR EXPERIMENTS FOR DETERMINATION OF HYDROGEN BONDS, DISTANCE RESTRAINTS AND RDCS

Hydrogen bonds are established by through-hydrogen bond scalar couplings, using a TROSY-type experiment, the quantitative HNN-COSY (Dingley and Grzesiek, 1998). This experiment unambiguously identifies hydrogen-bonding schemes in G–C and A–U Watson–Crick base pairs and non-Watson–Crick base pairs that involve imino to nitrogen hydrogen bonds, like reverse A–U Hoogsten and head-to-head G–A base pairs. Previously, hydrogen-bonding schemes were only identified based on their characteristic NOE patterns in water-NOESY spectra (Heus and Pardi, 1991a).

All through-bond assignments are usually also complemented by through-space-based assignments using NOESY-type experiments, such as homonuclear 2D NOESY experiments both in D_2O and H_2O, and 3D ^{13}C- or ^{15}N-edited NOESY experiments. The best result for exchangeable protons is usually achieved using the 2D S-NOESY pulse sequence (Smallcombe, 1993), followed by WATERGATE (Piotto et al., 1992) and WET (Smallcombe et al., 1995) suppression schemes. For 100% D_2O samples, a gentle presaturation is sufficient to suppress the residual HDO-signal. A 3D ^{13}C-edited NOESY is best performed with 100% D_2O RNA samples, since water suppression could again lead to loss of RNA signals under and around the water. The only drawback compared to H_2O samples is that ^{13}C-resolved NOE information between exchangeable and nonexchangeable protons is thereby lost (see (Lukavsky and Puglisi, 2001)).

2D heteronuclear ^{13}C and ^{15}N correlation experiments also need to be optimized to take the specific NMR properties of RNA molecules into account. In ^{15}N correlation experiments, different types of magnetization transfer and water suppression techniques are used for the observation of exchanging imino and amino resonances. Best results for exchange-broadened imino resonances are obtained using the jump-return ^{15}N HMQC (Szewczak et al., 1993) followed by the 3–9–19 WATERGATE ^{15}N HSQC (Piotto et al., 1992). Both experiments work well for RNAs up to 100 kDa, but better signal to noise ratio is achieved using the jump-return ^{15}N HMQC. Efficient magnetization transfer for exchange-broadened amino groups can be achieved using a CPMG pulse train instead of the common INEPT transfer implemented in the 2D ^{15}N CPMG HSQC (Mulder et al., 1996). For larger RNA molecules, a 3–9–19 WATERGATE ^{15}N TROSY (Pervushin et al., 1997) is used, which selects the slowest relaxing component of a given N1/3–H1/3 multiplet and therefore yields narrower line-widths compared to the jump-return ^{15}N HMQC or ^{15}N HSQC. However, since only one component of the multiplet is selected, the sensitivity of the ^{15}N TROSY is lower than that of the ^{15}N HMQC or ^{15}N HSQC. Nevertheless, the ^{15}N TROSY is the method of choice for the measurement of RDCs of small and large RNAs. A minimum of four individual ^{15}N TROSY experiments is required to determine ^{15}N RDCs. First, J_{NH} couplings are measured in the absence of Pf1 phage by selecting either the upfield or downfield ^{1}H component of the two ^{15}N downfield doublet of the N1/3–H1/3 multiplet using the ^{15}N TROSY experiments. Upon addition of Pf1 phage (10–20mg/ml), the same two ^{15}N TROSY experiments are performed again to measure J_{NH} couplings, which now also contain a contribution from the ^{1}H–^{15}N dipolar interaction. Simply subtracting the J_{NH} values in the presence and absence of Pf1 phage yields the RDC value for a given imino proton (see also (Hansen et al., 2000; Lukavsky and Puglisi, 2005)).

The measurement of RDCs from ^{1}H–^{13}C one-bond interactions follows the same principle. Figure 5 shows an adenine C2–H2 cross peak from a spectrum acquired with a 2D t_1- and t_2-coupled ^{1}H–^{13}C HSQC experiment on a 77nt RNA. Once again, only the downfield components in the ^{13}C-dimension yield sufficiently sharp resonances, while the upfield components of the multiplet exhibit severe broadening (Figure 5). One-bond ^{1}H–^{13}C couplings are measured using a ^{13}C-version of the Weigelt–TROSY (Weigelt, 1998), which includes a constant time (CT) evolution period and allows selecting either the upfield or downfield ^{1}H component of the ^{13}C downfield doublet (see Figure 5). The CT frequency-editing period is required to eliminate one-bond ^{13}C–^{13}C couplings during t_1-evolution (17ms for base and 25ms for sugar carbons). In large RNAs, where short transverse relaxation times are encountered, these long CT-periods can lead to significant losses in signal intensity and thereby make measurement lengthy or even impossible. The implementation of RDCs into structure calculations is described in detail in (Lukavsky and Puglisi, 2005).

The NMR experiments for the assignment of spin systems as well as the NOESY-type and torsion angle experiments, I described above, worked well for us with RNAs up to 55nt in length. For larger RNAs, where increasing resonance overlap and slower tumbling times are encountered, unambiguous resonance assignments can only be obtained by dissecting the larger RNA into thermodynamically stable subdomains, preferably between 30 and 40nt, where the above experiments work best. The way in which this "divide and conquer" approach is used in the structure determination of large RNAs is discussed in the next section.

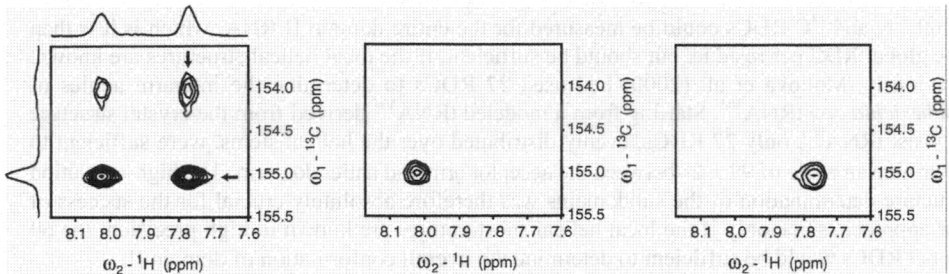

Figure 5. RDC-measurement using ^{1}H–^{13}C TROSY with CT evolution (Weigelt, 1998). Schematic representation of an adenine C2–H2 multiplet from a 2D t_{1}- and t_{2}-coupled ^{1}H–^{13}C HSQC experiment of the 77nt RNA comprising HCV IRES domain II (Lukavsky et al., 2003). The 2D ^{1}H–^{13}C ct-TROSY selects either the downfield or the upfield component of the downfield ^{13}C–^{1}H doublet.

3.4. STRUCTURE DETERMINATION OF LARGE RNAS WITH RDCS

High-resolution structure determination of a large, 25kDa RNA, by NMR spectroscopy, was first applied to domain II of the HCV IRES RNA. The basic idea, how to solve a 77nt RNA by NMR was as follows:

1. Divide the large RNA into thermodynamically stable subdomains of smaller size, where RNA NMR methods work efficiently and assignments are much easier.
2. Determine high-resolution structures of the subdomains using a maximum number of distance, torsion angle and RDC restraints (local).
3. Calculate structures of the entire large RNA using the distance, torsion angle and local RDC restraints from the subdomains AND global RDCs obtained from the large RNA to define the local and overall conformation with high precision.

The smaller RNA oligonucleotides comprising subdomains IIa (55nt) and IIb (34nt) of HCV IRES domain II (77nt) were carefully designed based on initial assignments of base pairs and helical stretches in the large RNA using a S-NOESY spectrum (Lukavsky and Puglisi, 2005). Comparison of cross-peak patterns and intensities in S-NOESY spectra of the subdomains and the large domain II RNA as well as comparison of chemical shifts in 2D HSQC spectra was used to confirm that the isolated subdomains fold autonomously into the same conformation as in the context of the large RNA. High-resolution structure determination of the subdomains employed the NMR techniques described in the previous chapter and included a refinement step with angular restraints derived from ^{13}C and ^{15}N RDCs using 136 (IIa) or 105 (IIb) RDCs, respectively. The final ensembles calculated with RDCs displayed an RMSD of 2.34 (IIa) and 1.35 (IIb) Å, respectively.

In order to define the global conformation, i.e. the relative orientation of the subdomains IIa and IIb in the context of the large RNA, RDCs were measured for the entire domain II RNA. Only imino N–H and aromatic C–H one-bond couplings could be measured reliably using the ^{15}N and ^{13}C TROSY experiments discussed above. Reliable measurement of ribose RDCs, on the other hand, was severely compromised by resonance overlap in the ribose region and significant signal loss during the 25ms CT-period in the ^{13}C TROSY experiments. We therefore decided to exclude ribose RDCs from the final refinement of the large RNA. A total of 60 ^{15}N and ^{13}C RDCs could be measured for the entire domain II RNA, which is less than one global RDC per residue, but should be sufficient, if the local helical structures are known. Previously, Mollova et al. (2000) had used 27 RDCs to determine the interarm angles of *Escherichia coli* tRNA^{Val}. Starting from a modeled tRNA^{Val} derived from the crystal structure of yeast tRNA^{Phe}, only 27 RDCs, evenly distributed over the helical stems, were sufficient to determine an angle of 99 ± 2° between the acceptor arm and anticodon arm. The high-resolution structure determination of the subdomains was therefore absolutely crucial for the success of our approach, since only if the local helical substructures are known to high precision, the 60 global RDCs would be sufficient to determine the overall conformation of domain II.

The final ensemble of domain II structures (Figure 6B) calculated with three individual sets of RDCs for domain IIa (112), IIb (89), and II (60), respectively, showed both locally improved

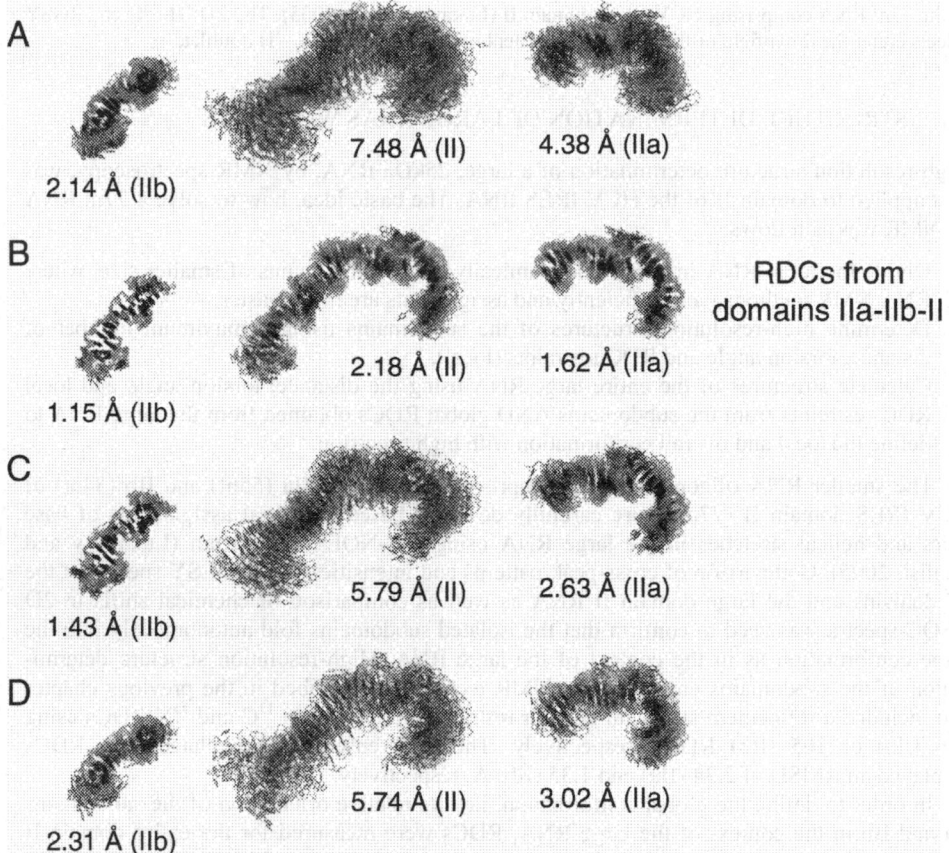

Figure 6. Final ensembles of HCV IRES domain II structures calculated with different sets of RDCs and corresponding local superpositions of subdomains IIa and IIb. (**A**) Final ensemble of structures calculated without RDCs. (**B**) Final ensemble of structures calculated with RDCs from domains IIa, IIb, and II. (**C**) Final ensemble of structures calculated only with RDCs from domains IIa and IIb. (**D**) Final ensemble of structures calculated only with RDCs from domain II.

subdomains (1.15 Å for IIb and 1.62 Å for IIa) and a well-defined global shape of domain II (2.18 Å) compared to the ensembles obtained without RDCs (Figure 6A) (RMSDs of 2.43 Å for IIb, 4.38 for IIa, and 7.48 Å for II). This significant improvement of the structural precision could only be achieved by the combination of local RDCs from the subdomains with the global RDCs from domain II. Omitting the 60 RDCs from domain II yielded well-defined structures for the local domains (1.43 Å for IIb and 2.63 Å for IIa), while the overall shape of domain II improved only slightly (5.79 Å) as shown in Figure 6C. Similarly, using only 60 RDCs from domain II also gave only a slight improvement of the overall definition (5.74 Å), since subdomains were less well-defined without RDCs (2.31 Å for IIb and 3.02 Å for IIa) as shown in Figure 6D. A de novo structure determination of a large RNA, where local and global structures are unknown, therefore requires both local and global RDCs to define both the local subdomains and the overall conformation with high precision.

3.5. FUTURE PERSPECTIVES FOR RNA NMR SPECTROSCOPY

The protocol described above should allow determining structures of even larger RNAs by NMR spectroscopy, because a larger RNA (up to 150nt) could be simply dissected into more than two thermodynamically stable subdomains to obtain local, high-precision structures. The major limitation of this approach will be the slow molecular tumbling times of very large, extended RNAs and the concomitant short transverse relaxation times. This will lead to increased resonance linewidths and make it impossible to obtain reliable global RDCs. On the other hand, structure determination of a more compactly folded 150nt RNA, which should not tumble as slowly as an extended 150nt RNA, could be feasible using our approach.

Another limitation arises from the ever-increasing resonance overlap problem encountered with the entire large RNA, which will make the extraction of sufficient global RDCs difficult to impossible. Nucleoside-type specific isotopic labeling of the large RNA could help to alleviate this problem to a certain extent. But in our experience, resonance overlap frequently occurs between different residues of the same nucleoside-type, i.e. adenosine resonances overlap with other adenosines, uridine resonances with other uridines, because they reside in the same chemical environment within the large RNA, which gives rise to very similar or the same chemical shifts. In my laboratory, we recently encountered such a problem during the structure determination of a 74nt RNA comprising the dendritic targeting element of brain cytoplasmic 1 (BC1) RNA (Tiedge et al., 1993). Severe resonance overlap between residues of the same nucleoside-type (especially in helical regions) significantly reduced the number of obtainable global RDCs and could only be alleviated using complementary, segmentally labeled RNAs prepared by enzymatic ligation (Tzakos et al., 2006a). This methodology requires two individual NMR samples of the large RNA with complementary isotopic labeling schemes. In this specific case, one sample was prepared, in which part of the RNA is ^{15}N-labeled and the other part ^{13}C-labeled and one with the complementary isotopic labels (Tzakos et al., 2006a). Compared to the uniformly ^{13}C,[15]N-labeled sample of this large RNA, double the number of global RDCs could be obtained using the complementary labeled samples, which greatly benefited the precision of the global RNA structure (Tzakos et al., 2006c). If this approach is combined with single nucleoside-type specific labeling or even combined with partial deuteration, sufficient global RDCs could be obtained even for very large RNAs and permit their structure determination by NMR spectroscopy.

4. Acknowledgment

I would like to thank Andreas G. Tzakos for his careful reading and Manolia Margaris for her final editing of the manuscript.

5. References

1. Allain, F.H. and Varani, G. (1997). How accurately and precisely can RNA structure be determined by NMR? *J. Mol. Biol.*, 267, 338–351.
2. Bax, A., Kontaxis, G., and Tjandra, N. (2001). Dipolar couplings in macromolecular structure determination. *Methods Enzymol.* 339, 127–174.
3. Cheong, C., Varani, G., and Tinoco, I., Jr. (1990). Solution structure of an unusually stable RNA hairpin, 5'GGAC(UUCG)GUCC [see comments]. *Nature*, 346, 680–682.
4. Davis, J.H., Tonelli, M., Scott, L.G., Jaeger, L., Williamson, J.R., and Butcher, S.E. (2005). RNA helical packing in solution: NMR structure of a 30 kDa GAAA tetraloop-receptor complex. *J. Mol. Biol.* 351, 371–382.
5. Dingley, A.J. and Grzesiek, S. (1998). Direct observation of hydrogen bonds in nucleic acid base pairs by internucleotide $^2J_{NN}$ couplings. *J. Am. Chem. Soc.* 120, 8293–8297.
6. D'Souza, V. and Summers, M.F. (2004). Structural basis for packaging the dimeric genome of Moloney murine leukaemia virus. *Nature*, 431, 586–590.
7. Fiala, R., Czernek, J., and Sklenar, V. (2000). Transverse relaxation optimized triple-resonance NMR experiments for nucleic acids. *J. Biomol. NMR*, 16, 291–302.
8. Hansen, M.R., Mueller, L., and Pardi, A. (1998). Tunable alignment of macromolecules by filamentous phage yields dipolar coupling interactions. *Nat. Struct. Biol.* 5, 1065–1074.
9. Hansen, M.R., Simorre, J.P., Hanson, P., Mokler, V., Bellon, L., Beigelman, L., and Pardi, A. (1999). Identification and characterization of a novel high affinity metal-binding site in the hammerhead ribozyme. *RNA*, 5, 1099–1104.
10. Hansen, M.R., Hanson, P., and Pardi, A. (2000). Filamentous bacteriophage for aligning RNA, DNA, and proteins for measurement of nuclear magnetic resonance dipolar coupling interactions. *Methods Enzymol.* 317, 220–240
11. Heus, H.A. and Pardi, A. (1991a). Novel H-1 nuclear magnetic resonance assignment procedure for RNA duplexes. *J. Am. Chem. Soc.* 113, 4360–4361.
12. Heus, H.A. and Pardi, A. (1991b). Structural features that give rise to the unusual stability of RNA hairpins containing GNRA loops. *Science*, 253, 191–194.
13. Kay, L.E., Ikura, M., and Bax, A. (1990). Proton–proton correlation via carbon carbon couplings – a 3-dimensional NMR approach for the assignment of aliphatic resonances in proteins labeled with carbon-13. *J. Am. Chem. Soc.* 112, 888–889.
14. Kay, L.E., Xu, G.Y., Singer, A.U., Muhandiram, D.R., and Formankay, J.D. (1993). A gradient-enhanced Hcch-Tocsy experiment for recording side-chain H-1 and C-13 correlations in H2O samples of proteins. *J. Magn. Reson. Series B*, 101, 333–337.
15. Kim, I., Lukavsky, P.J., and Puglisi, J.D. (2002). NMR study of 100 kDa HCV IRES RNA using segmental isotope labeling. *J. Am. Chem. Soc.* 124, 9338–9339.
16. Latham, M.P., Brown, D.J., McCallum, S.A., and Pardi, A. (2005). NMR methods for studying the structure and dynamics of RNA. *Chembiochem.* 6, 1492–1505.
17. Lukavsky, P.J., Otto, G.A., Lancaster, A.M., Sarnow, P., and Puglisi, J.D. (2000). Structures of two RNA domains essential for hepatitis C virus internal ribosome entry site function. *Nat. Struct. Biol.* 7, 1105–1110.
18. Lukavsky, P.J. and Puglisi, J.D. (2001). RNAPack: an integrated NMR approach to RNA structure determination. *Methods*, 25, 316–332.

19. Lukavsky, P.J., Kim, I., Otto, G.A., and Puglisi, J.D. (2003). Structure of HCV IRES domain II determined by NMR. *Nat. Struct. Biol.* 10, 1033–1038.
20. Lukavsky, P.J. and Puglisi, J.D. (2005). Structure determination of large biological RNAs. *Methods Enzymol.* 394, 399–416.
21. Marino, J.P., Prestegard, J.H., and Crothers, D.M. (1994a). Correlation of adenine H2/H8 resonances in uniformly C-13 labeled RNAs by 2d Hcch-Tocsy – a new tool for H-1 assignment. *J. Am. Chem. Soc.* 116, 2205–2206.
22. Marino, J.P., Schwalbe, H., Anklin, C., Bermel, W., Crothers, D.M., and Griesinger, C. (1994b). A three-dimensional triple-resonance H-1,C-13,P-31 experiment – sequential through-bond correlation of ribose protons and intervening phosphorus along the RNA oligonucleotide backbone. *J. Am. Chem. Soc.* 116, 6472–6473.
23. Marino, J.P., Diener, J.L., Moore, P.B., and Griesinger, C. (1997). Multiple-quantum coherence dramatically enhances the sensitivity of CH and CH2 correlations in uniformly C-13-labeled RNA. *J. Am. Chem. Soc.* 119, 7361–7366.
24. Marino, J.P., Schwalbe, H., and Griesinger, C. (1999). J-coupling restraints in RNA structure determination. *Accounts Chem. Res.* 32, 614–623.
25. Mollova, E., Hansen, M.R., and Pardi, A. (2000). Global structure of RNA determined with residual dipolar couplings. *J. Am. Chem. Soc.* 122, 11561–11562.
26. Mollova, E.T. and Pardi, A. (2000). NMR solution structure determination of RNAs. *Curr. Opin. Struc. Biol.* 10, 298–302.
27. Mulder, F.A.A., Spronk, C., Slijper, M., Kaptein, R., and Boelens, R. (1996). Improved Hsqc experiments for the observation of exchange broadened signals. *J. Biomol. NMR*, 8, 223–228.
28. Pardi, A. and Nikonowicz, E.P. (1992). Simple procedure for resonance assignment of the sugar protons in C-13-labeled RNAs. *J. Am. Chem. Soc.* 114, 9202–9203.
29. Pervushin, K., Riek, R., Wider, G., and Wuthrich, K. (1997). Attenuated T2 relaxation by mutual cancellation of dipole-dipole coupling and chemical shift anisotropy indicates an avenue to NMR structures of very large biological macromolecules in solution. *Proc. Natl. Acad. Sci. U S A*, 94, 12366–12371.
30. Piotto, M., Saudek, V., and Sklenar, V. (1992). Gradient-tailored excitation for single-quantum NMR spectroscopy of aqueous solutions. *J. Biomol. NMR*, 2, 661–665.
31. Scott, L.G., Tolbert, T.J., and Williamson, J.R. (2000). Preparation of specifically 2H- and 13C-labeled ribonucleotides. *Methods Enzymol.* 317, 18–38.
32. Simon, B., Zanier, K., and Sattler, M. (2001). A TROSY relayed HCCH-COSY experiment for correlating adenine H2/H8 resonances in uniformly 13C-labeled RNA molecules. *J. Biomol. NMR*, 20, 173–176.
33. Simorre, J.P., Zimmermann, G.R., Pardi, A., Farmer, B.T., and Mueller, L. (1995). Triple resonance Hnccch experiments for correlating exchangeable and nonexchangeable cytidine and uridine base protons in RNA. *J. Biomol. NMR*, 6, 427–432.
34. Simorre, J.P., Zimmermann, G.R., Mueller, L., and Pardi, A. (1996a). Correlation of the guanosine exchangeable and nonexchangeable base protons In C-13-(In)-N-15-labeled RNA with an Hnc-Tocsy-Ch experiment. *J. Biomol. NMR*, 7, 153–156.
35. Simorre, J.P., Zimmermann, G.R., Mueller, L., and Pardi, A. (1996b). Triple-resonance experiments for assignment of adenine base resonances in C-13/N-15-labeled RNA. *J. Am. Chem. Soc.* 118, 5316–5317.
36. Sklenar, V., Miyashiro, H., Zon, G., Miles, H.T., and Bax, A. (1986). Assignment of the 31P and 1H resonances in oligonucleotides by two-dimensional NMR spectroscopy. *Febs. Lett.* 208, 94–98.
37. Smallcombe, S.H. (1993). Solvent suppression with symmetrically-shifted pulses. *J. Am. Chem. Soc.* 115, 4776–4785.
38. Smallcombe, S.H., Patt, S.L., and Keifer, P.A. (1995). Wet solvent uppression and Its Applications to Lc Nmr and High-Resolution NMR spectroscopy. *J. Mag. Res. Series A*, 117, 295–303.

39. Szewczak, A.A., Kellogg, G.W., and Moore, P.B. (1993). Assignment of NH resonances in nucleic acids using natural abundance 15N-1H correlation spectroscopy with spin-echo and gradient pulses. *FEBS Lett.* 327, 261–264.
40. Tiedge, H., Chen, W., Brosius, J., and Fishberg Research Center for Neurobiology, M.S.S.o.M.N.Y.N.Y. (1993). Primary structure, neural-specific expression, and dendritic location of human BC200 RNA. *J. Neurosci.* 13(6), 2382–2390.
41. Tjandra, N. and Bax, A. (1997). Direct measurement of distances and angles in biomolecules by NMR in a dilute liquid crystalline medium. *Science,* 278, 1111–1114.
42. Tzakos, A.G., Easton, L.E., and Lukavsky, P.J. (2006a). Complementary segmental labeling of large RNAs: economic preparation and simplified NMR spectra for measurement of more RDCs. *J. Am. Chem. Soc.* 128, 13344–13345.
43. Tzakos, A.G., Grace, C.R., Lukavsky, P.J., and Riek, R. (2006b). NMR techniques for very large proteins and rnas in solution. *Annu. Rev. Biophys. Biomol. Struct.* 35, 319–342.
44. Tzakos, A.G. et al. (2006c). unpublished results.
45. Varani, G., Aboulela, F., and Allain, F.H.T. (1996). NMR investigation of RNA structure. *Prog. Nucl. Mag. Res. Sp.* 29, 51–127.
46. Warren, J.J. and Moore, P.B. (2001). Application of dipolar coupling data to the refinement of the solution structure of the sarcin-ricin loop RNA. *J. Biomol. NMR,* 20, 311–323.
47. Weigelt, J. (1998). Single scan, sensitivity- and gradient-enhanced TROSY for multidimensional NMR experiments. *J. Am. Chem. Soc.* 120, 10778–10779.

PROBABILISTIC STRUCTURE CALCULATION

WOLFGANG RIEPING,[1,*] MICHAEL HABECK,[2,*] AND MICHAEL NILGES
Unité de Bio-Informatique Structurale, Institut Pasteur and CNRS URA 2185, 25–28 rue du docteur Roux, F-75015 Paris, France
[1]*Current address: Department of Biochemistry, University of Cambridge, 80 Tennis Court Road, Cambridge CB2 1GA, UK*
[2]*Current address: Max Planck Institute for Developmental Biology, Spemannstrasse 35 and Max Planck Institute for Biological Cybernetics, Spemannstrasse 38, 72076 Tübingen, Germany*
[*] These authors contributed equally to this work

1.   Overview: structure determination, a problem without unique solution

The calculation of three-dimensional structures from data relies on a quantitative model that relates the data to molecular conformation. A so-called *forward model* describes the results of measurements as a function of the atomic coordinates. At first glance it would seem that an ideal structure determination algorithm should be implementing the inverse of the forward model and evaluate it for the measurements to obtain the structure. Strictly speaking, such an inversion is impossible for several reasons. First, if different conformations result in the same data, the forward model is inherently degenerate, meaning that several models exist that explain the data equally well. Second, even a formally invertable model is degenerate in practice if the data are incomplete. Third, both data and forward model are uncertain: measurements are subject to errors, theories rely on approximations. Finally, the forward model usually involves auxiliary parameters that are not measurable.

A simple example illustrates some of these points. Three-bond scalar couplings $^3J(\varphi)$ are related to molecular conformation (the torsion angle between the first and the third bond, φ) through the well-known Karplus relation [1]. The forward model in this case is the equation

$$^3J(\varphi) = A\cos^2\varphi + B\cos\varphi + C. \tag{1}$$

Figure 1 shows, with a hypothetical Karplus curve, the effects of small errors in the data (small variations in the measured coupling constants). A small difference in the measured coupling leads to a large difference in the resulting angle ϕ.

Figure 2 shows, again with a hypothetical Karplus curve, the effects of small variations in the coefficients A, B, and C (i.e. small variations in the forward model). Again, small variations lead to a potentially large difference of the interpretation of the coupling constant in terms of structure.

J. D. Puglisi (ed.), Structure and Biophysics – New Technologies for Current Challenges in Biology and Beyond, 81–98.
© 2007 *Springer.*

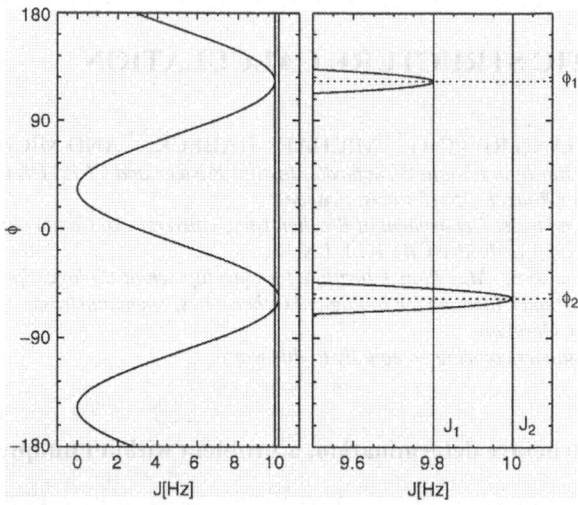

Figure 1. Hypothetical Karplus curve to illustrate the influence of small errors in the data on structural interpretation. The left panel shows the whole curve, the right a magnification of the region around 10 Hz.

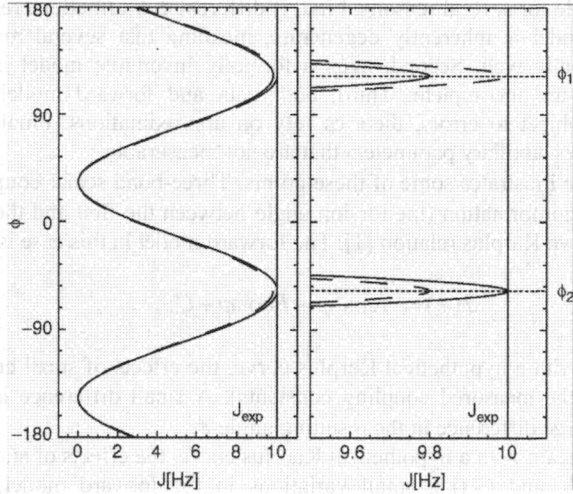

Figure 2. Hypothetical Karplus curve to illustrate the influence of small variations in the theory on structural interpretation. The left panel shows the whole curve, the right a magnification of the region around 10 Hz.

In other words, the structure can be highly sensitive against inaccuracies in the data and variations in the forward model. As the magnitude of both components is a priori unknown, the major challenge for structure calculation techniques is to deal with such uncertainties.

Despite the general awareness of these problems, the current paradigm of structure calculation is to attempt a direct inversion of the data. Most algorithms minimize a hybrid energy $E_{phys} + w_{data} E_{data}$, where a non-physical term E_{data} assesses the match between data and structure (an example is a least-squares function such as a harmonic potential). A physical energy E_{phys} serves as a regulariser and partially removes the degeneracy of the problem (e.g. known covalent bond lengths and bond angles are used to complement the experimental data). The rationale is that minimization of the hybrid energy effectively inverts the forward model yielding the "true" structure. The problems illustrated above are treated by heuristic methods, for example, by deriving bounds on torsion angles [2, 3] rather than giving a precise angle. These bounds need to be chosen large enough to encompass errors introduced both by the measurement (experimental error) and the forward model (errors introduced by variations in the parameters A, B, and C; influence of factors not incorporated in the simple analytical form of the Karplus relationship). However, using bounds leads to a loss of experimental information, and to a more degenerate forward model (many conformations can lie between the bounds).

This strategy works in the case of many data of good quality. Even then, it is however difficult to assess the reliability of the result, and the dependency of the result on the parameters chosen for the calculation (in the case of torsion angles, the width of the bounds, and the influence of the parameters A, B, and C). In less favourable situations, structure calculation by inversion may pose difficulties. Specifically, it remains unclear how to choose auxiliary parameters like w_{data} (the weight of the data relative to the physical energy) and unknown parameters of the forward model (as, for instance, A, B, and C). The minimization paradigm offers no general principle for their estimation. Instead, heuristics and empirical rules need to be introduced that contain ad hoc elements and are not guaranteed to be consistent. As a result, the derived coordinates are sensitive to small changes in the data and to the choice of auxiliary parameters.

1.1. STRUCTURE DETERMINATION BY INFERENCE

These problems can be avoided if one considers structure determination as a problem of *inference* (rather than *deduction*). If a hypothesis must be assessed upon insufficient evidence, deductive reasoning breaks down, and one needs to restrict oneself to inductive inferences. In order to make quantitative statements, we assign a number P_i to each possible conformation X_i. P_i assesses the plausibility that the molecular structure is X_i, $P_i > P_j$ meaning that conformation X_i is more plausible than X_j. Because the X_i are continuous, a continuum of P_i values is distributed over conformational space; the distribution reflects the extent to which an inversion is possible. If all but one P_i vanish, the data are uniquely invertible. Thus, inference contains deduction as a limiting case. In the case of uniform P_i, the data are completely uninformative with respect to the structure.

We illustrate this in Figure 3, again for the case of torsion angles and coupling constants. The figure shows a Karplus curve, with shading indicating the plausibility that a measured coupling agrees the value predicted by a forward model. Rather than trying to determine

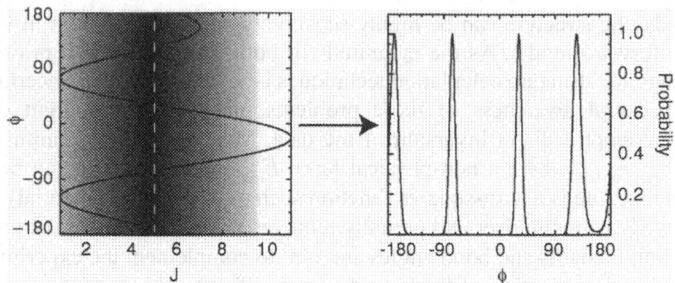

Figure 3. The left panel shows the Karplus curve for a particular set of values for *A*, *B*, and *C*. The *dashed line* signifies the measured coupling. The *shading* indicate the plausibility of values of *J* reproduced by a forward model: values close to the experimental value are more plausible than values far away. This leads to an assignment of a probability to different values of ϕ, indicated in the right panel.

which angle produced the experimentally observed coupling, we only say that structures that, according to a given theory, reproduce the coupling better are more plausible, others less. This leads to a distribution of plausibilities, shown on the right hand. This plausibility is equivalent to a probability [4] and necessarily follows the rules of probability theory. To solve inference problems quantitatively, we therefore need to employ probability theory. The framework of inferential structure determination (ISD) [5] states that any structure determination problem amounts to calculating the probabilities $\{P_1, P_2, ...\}$ for every conformation of the structure.

We demand the probabilities P_i to be "objective" in the sense that they should only depend on the data D and on relevant background information I needed to analyze the experiment (such as the forward model or knowledge about physical interactions). Thus P_i is a conditional probability, $P_i = P(X_i \mid D, I)$. This conditional probability is not a frequency of occurence, but a quantitative representation of our state of knowledge. In the case of a continuous parametrization (Cartesian coordinates, dihedral angles) P_i is a density $p(X \mid D, I)$.

Calculation of this probability density is possible by using Bayes's theorem [6], a direct consequence of probability calculus:

$$p(X \mid D, I) \propto \pi(X \mid I) L(D \mid X, I). \tag{2}$$

The so-called posterior distribution $p(X \mid D, I)$ factorizes into two natural component: The prior distribution $\pi(X \mid I)$ describes our knowledge about general properties of biomolecular structures in the absence of any data. The likelihood function $L(D \mid X, I)$ quantifies the probability of observing the data D given a molecular structure X. Hence, L is a function of the coordinates and is composed of a forward model to predict the measurements, and an error distribution to account for deviations of measured from calculated values.

Probability theory formally solves our inference problem and avoids difficulties that occur with a direct inversion. The forward model is contained in the likelihood; errors in the model or in the data are modelled explicitly. If the model is exactly invertible, the likelihood approaches a function that is sharply peaked around the inverse model evaluated at the data. The amount and the quality of the data determine the precision of the coordinates. The prior takes additional knowledge about the structure into account without bias.

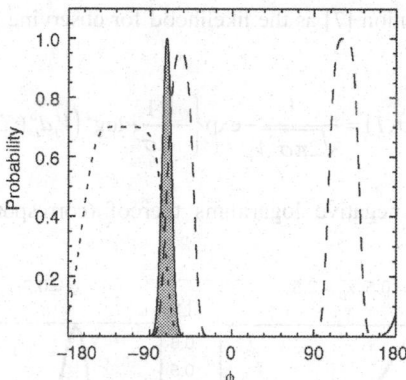

Figure 4. Prior knowledge is incorporated in a natural way using the laws of probability theory. In the illustrated case, the prior knowledge (*dotted line*) is the probability to observe a particular torsion angle before any data are measured (for example, we know that the protein backbone torsion angle ϕ is in most cases negative). The likelihood (*dashed line*) adds the knowledge obtained from the data: in our case, there are two peaks in the likelihood. The posterior probability (*solid line*) is obtained by multiplication of the prior probability and the likelihood and represents the total knowledge we have about the conformation.

1.2.   THE ERROR MODEL AND THE LIKELIHOOD

For coupling constants, the least informative (in the sense of information theory) error model assuming no systematic deviation and knowledge of the average discrepancy σ is described by a Gaussian [6]:

$$p(^3J\,|\,\varphi,A,B,C,\sigma)=\frac{1}{\sqrt{2\pi\sigma^2}}\exp\left\{-\frac{1}{2}\left(^3J-^3J(\varphi)\right)^2\right\} \qquad (3)$$

In the case of NOE measurements, the data consist of a list of assigned cross-peaks with volumes or intensities V_i, i.e. $D=\{V_1,...,V_n\}$. When we use the ISPA (Isolated Spin Pair Approximation) to calculate the volume of a cross-peak:

$$V_i=\gamma d_i^{-6}(X). \qquad (4)$$

we introduce an unknown parameter, the scale ("calibration factor") γ. Here, $d_i(X)$ denotes the distance in the structure X that corresponds to volume V_i. Due to imperfections in the data (e.g. errors during data processing, experimental noise) and the imperfections of ISPA (e.g. effects of dynamics and spin diffusion are not taken into account), measured volumes deviate from the theoretical predictions. We introduce the parameter σ which quantifies the discrepancy between measured and calculated volumes. That is, σ collectively accounts for the error made by the ISPA and the experimental noise. Taking into account the positivity of volumes, we use a lognormal distribution [7] as the likelihood for observing a cross-peak with volume V_i

$$p(V_i \mid X, \gamma, \sigma, I) = \frac{1}{\sqrt{2\pi\sigma^2}V_i} \exp\left\{-\frac{1}{2\sigma^2}\log^2\left(V_i d_i^6(X)/\gamma\right)\right\}. \tag{5}$$

This distribution and the negative logarithms thereof (corresponding to an energy) are depicted in Figure 5.

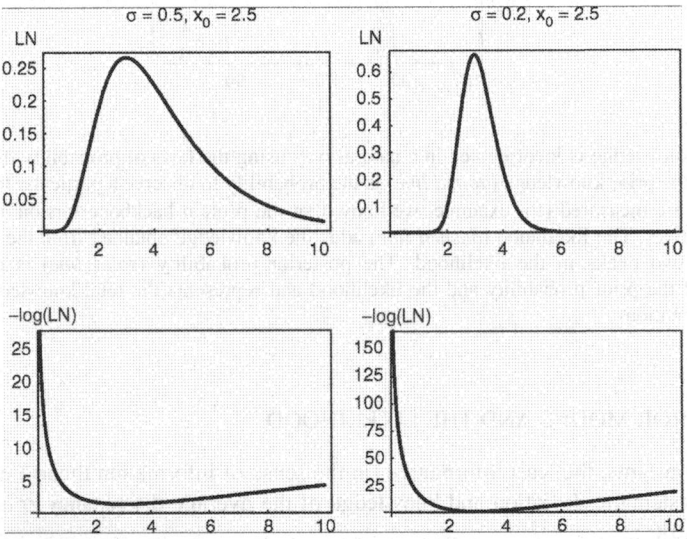

Figure 5. Lognormal distribution for two different values of σ, and the negative logarithm of the distribution, corresponding to an energy. Note the asymmetry of the distribution.

1.3. NUISANCE PARAMETERS

As we have seen, it is often necessary to introduce auxiliary parameters in order to describe a structure determination problem adequately. For example, the coefficients A, B, C of the Karplus relationship are, strictly speaking, unknown for the particular protein that one is investigating. Also, the data quality is an unknown parameter, and the calibration factor for NOE volumes. Obviously, empirical rules can be used to estimate these parameters, but one of the decisive advantages of a Bayesian approach to structure determination is the treatment of the unknown parameters. All additional unknown parameters, which we call $\xi_i, i = 1, N_\xi$, of the error model and of the theory are estimated along with the structure. In Bayesian theory, these parameters are "nuisance parameters" that are treated in the same way as the coordinates. We simply replace X with (X, ξ) in Eq. (2), and the full posterior becomes

$$p(X,\xi \mid D,I) \propto \pi(X \mid I)\pi(\xi \mid I)L(D \mid X,\xi,I). \tag{6}$$

Here, we a priori assumed independence of X and ξ – our prior knowledge about the coordinates is independent of that about the nuisance parameters ξ_i – and we introduced the additional prior distribution $\pi(\xi \mid I)$ (Jeffreys' prior [8] in our applications).

As indicated by the joint distribution in Eq. (6), the values of the nuisance parameters ξ are uncertain for the very same reason the probability of the coordinates is distributed: our information at hand is insufficient to determine them uniquely. In order to account for our ignorance regarding the nuisance parameters, we integrate the joint posterior distribution over ξ (also called marginalization [6]):

$$p(X \mid D,I) = \int d\xi p(X,\xi \mid D,I) \propto p(X \mid I) \int d\xi p(D \mid X,\xi,I)p(\xi \mid I). \tag{7}$$

That is, we obtain the posterior distribution for the coordinates by themselves by replacing $p(D \mid X,I)$ in Eq. (2) with a weighted average over the likelihood conditioned on all possible values of all ξ_i. This is in marked difference to standard structure determination by minimization, where the value of any unknown parameter needs to be determined beforehand by empirical rules and held constant during the calculation, and therfore only one single value is used. In contrast, the result of a structure calculation by inference directly contains the influence of the uncertainty of the additional parameters.

2. Sampling of probability distributions

2.1. SAMPLING VS. MINIMIZATION

For a single – or very few – degrees of freedom, one could calculate the probability of every conformation for example by a grid search. For a larger number of degrees of freedom this is not possible and the space of possible conformations has to be explored by a suitable sampling algorithm. A good sampling algorithm will produce samples with the correct probability. That is, the probability can be directly calculated from the number of times a particular region is visited. This is illustrated in Figure 6.

The ensembles of NMR structures generated by repeated empirical minimization of the hybrid energy is often called "sampling". The minimization of a hybrid energy is equivalent to maximization of the probability with fixed nuisance parameters, since the hybrid energy can be interpreted as the negative logarithm of a probability. However, repeated maximization of the probability is not sampling since it does not produce samples with the correct probability. This can be easily seen from a very simple example, a probability surface with one unique maximum, equivalent to a single minimum in the hybrid energy. The minimum can be calculated by standard minimization algorithms, a simple task since there is no "multiple minimum problem". Repeated minimization will find exactly the same minimum in every calculation and will therefore not be able to give us any information about the shape of the hybrid energy around its minimum, or, equivalently, of the probability distribution around the maximum.

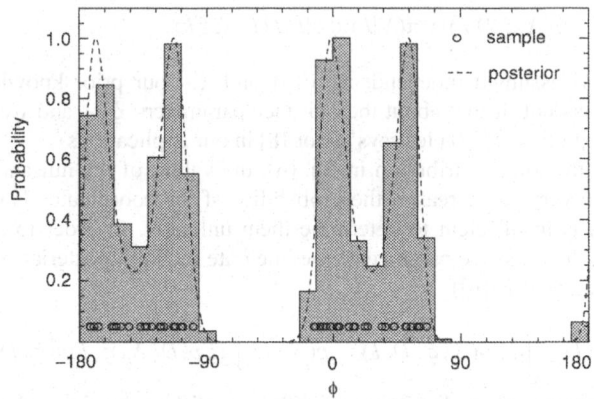

Figure 6. Illustration of exploration of probability distributions by sampling. An ideal sampling algorithm generates conformations (indicated by *small circles*) with the correct probability. For example, there are no or very few conformations in the low probability region in the middle. The histogram (indicated by *grey* surface) calculated from the samples approximates the true probability distribution.

2.2. MARKOV CHAIN MONTE CARLO SAMPLING WITH AN EXTENDED REPLICA–EXCHANGE METHOD

Sampling of the probability distribution of protein conformations is very difficult, due to a number of factors: there are many degrees of freedom (essentially the main chain torsion angles of a protein); the degrees of freedom are highly coupled; islands of high probability are separated by large stretches of low probability. The difficulties are related to those encountered in minimization, but are increased due to the fact that we not only look for the minimum of an energy (equivalent to the maximum probability) but try to determine the whole probability density in the relevant regions. The sampling algorithm needs to generate samples of protein conformation in the high probability regions, and needs to be able to get out of one high probability region to sample also the other ones. In minimization, trapping in local energy minima is avoided by using simulated annealing approaches. We cannot employ them for sampling since they are designed to search for a minimum energy and therefore do not generate samples characterizing a probability surface.

To address this problem, we proposed an extended replica-exchange Monte Carlo scheme for simulating the posterior densities [9]. Our sampling scheme uses three concepts to draw a sample from the joint posterior distribtion: a replica scheme, with a generalization of the usually employed temperature to other parameters; and two different Monte Carlo sampling algorithms to generate random samples for the coordinates (hybrid Monte Carlo (HMC)) and the nuisance parameters (Gibbs sampling).

A replica-exchange algorithm does not generate a single Markov chain but several chains in parallel (in our application, 50). Each of these 50 copies is run at different conditions (usually at different temperatures). At regular intervals, the chains are coupled by interchanging conformations from neighbouring copies. The exchange probabilites are adjusted in such a way that the detailed balance is preserved, an essential requirement to ensure unbiased sampling. This allows a trajectory being trapped in a particular conformation to escape through steps at higher temperatures. The concept is illustrated schematically in Figure 7.

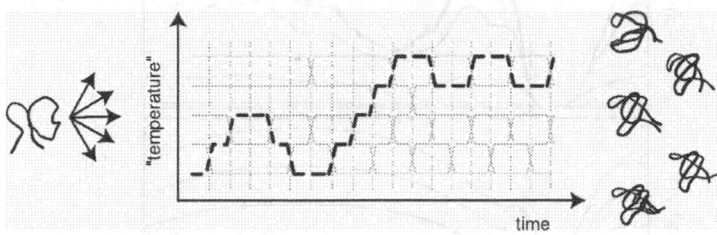

Figure 7. Schematic of a replica-exchange calculation. Several trajectories are run in parallel at different conditions. At regular intervals, the conformations of two neighbouring trajectories can interchange.

The generalization of the usual idea consists of introducing two parameters, q and λ, which have a similar effect as raising the temperature but allow a more tailored sampling. Here, q and λ control the shape of the distribution. For the "low-temperature distribution" (which marks the left side of the replica arrangement) these two parameters are chosen such, as to match the target distribution; for the "high temperature distribution", they have a similar effect as a high temperature does (i.e. "high energy" configurations become more likely). The sets of values for q and λ need to be chosen in such a way it is easy to sample configurations from the "high-temperature distribution".

The two parameters λ and q control independently the two factors which make up the posterior density:

$$f(X;\lambda,q) = \left[L(X)\right]^{\lambda} \pi(X;q). \tag{8}$$

where $f(X;\lambda,q)$ is a function of the posterior density (depending on the coordinates X) and the two parameters. The parameter λ modifies the weight on the likelihood function and thus determines the influence of the data. The extreme values are $\lambda = 1$ (the data are fully taken into account) and $\lambda = 0$ (the data are completely ignored).

We use the parameter q to replace the canonical ensemble $\pi(X)$ by a Tsallis generalized ensemble [10] $\pi(X;q) \propto \exp\{-\beta E(X;q)\}$. q parametrizes a non-linear transformation of the potential energy:

$$E(X;q) = \frac{q}{\beta(q-1)} \log\{q-1)(E(X) - E_{min})\} \tag{9}$$

where E_{min} is chosen such that $E(X) \geq E_{min}$ holds for all configurations X. For $E(X) > E_{min}$ and $q > 1$ the transformed energy becomes smoother, which enhances the mobility of the Markov chain. In the low energy regime $\beta(q-1)(E(X) - E_{min}) << 1$ the Tsallis ensemble reduces to the Boltzmann ensemble. In particular it holds that $E(X;1) = E(X) - E_{min}$.

The effect of the two replica parameters on the potential energy and on the likelihood function is illustrated in Figure 8.

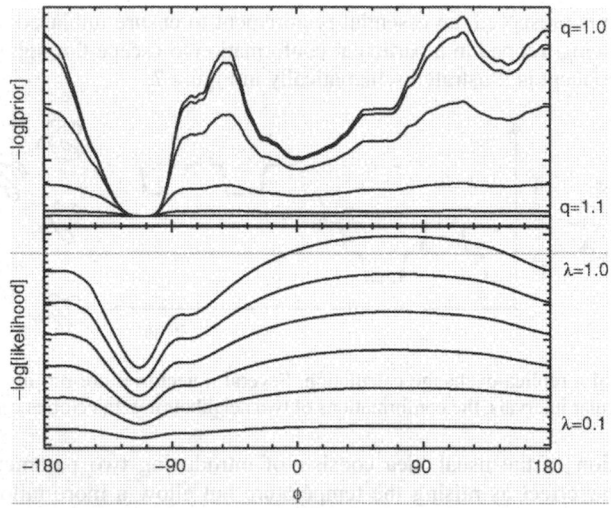

Figure 8. Effect of the replica parameters q and λ on the prior density and the likelihood function, respectively. In the upper panel, the effective potential energy $E(\theta;q)$ is shown for $q = 1 + 10^{-\alpha}$ varying from 1.0 ($\alpha = \infty$) to 1.001 ($\alpha = 3$). The lower panel shows the effect of weighing the likelihood function by letting λ range from 1.0 to 0.1. The curves are obtained by varying the torsion angle φ in the amino acid Alanine-122 of a Fyn SH3 domain in a folded configuration.

2.3.   SIMULTANOUS SAMPLING OF COORDINATES AND NUISANCE PARAMETERS

2.3.1. Gibbs sampling

The Gibbs algorithm [11] provides a general scheme for the simulation of probability distributions with conditional posteriors (i.e. posteriors for one variable given values for all the others) that are easy to simulate. The variables are updated in an iterative fashion by successively drawing samples from the conditional posteriors while fixing the other variables to their most recent values. The posterior can be decomposed into conditional posteriors for the error σ, the scale γ and coordinates X [7, 12]. The error σ and the scale γ can be directly sampled with random number generators. The conformational conditional posterior, $p(X|\sigma,\gamma,D,I)$, exhibits a much more complex topology and cannot be sampled directly.

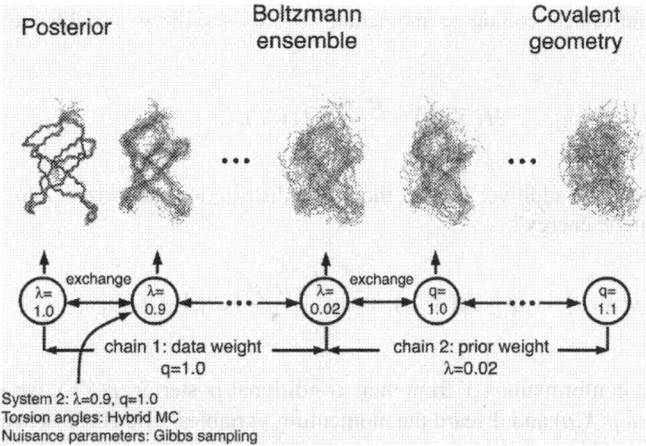

Posterior   Boltzmann   Covalent
               ensemble   geometry

chain 1: data weight
q=1.0

chain 2: prior weight
λ=0.02

System 2: λ=0.9, q=1.0
Torsion angles: Hybrid MC
Nuisance parameters: Gibbs sampling

Figure 9. Overall sampling strategy, using the replica algorithm, hybrid Monte Carlo to sample the coordinates, and Gibbs sampling to sample nuisance parameters. There are two chains separately for the two replica parameters λ and q. At a "high temperature" (right-hand side of the diagram), both the prior information is almost switched off using the parameter q, and the likelihood is strongly weighted down using λ. The only remaining information is therefore contained in the covalent geometry (inherently present due to the choice of torsion angles as parameters). In the middle, the prior information is fully present, and at the left side of the diagram, the "low temperature", both data and prior information are fully used. For each conformation in each replica, the nuisance parameters are sampled with Gibbs sampling.

2.3.2. Hybrid Monte Carlo

In order to deal with the high dimensionality and correlation of the posterior for the molecular conformations, we use the HMC method [13]. The key idea is to combine Metropolis Monte Carlo [14] and molecular dynamics (MD) to deal with correlated variables and to produce non-local proposal conformations while maintaining high acceptance rates. HMC is not restricted to molecular coordinates but can be used for any non-zero and differentiable probability distribution.

In order to use MD we reformulate the problem in physical terms: First, we write the conditional posterior as $p(X \mid \supseteq) \propto \exp\{-U(X)\}$, which defines an effective "potential energy"-function, similar to the hybrid energy function used in structure calculation by minimization. This effective potential energy needs to be defined for algorithmic purposes and has no "physical" significance.

Second, we introduce conjugate momenta, μ_i, as auxiliary variables to define the Hamiltonian

$$H(X,\mu) = \frac{\mu^T \mu}{2} + U(X).$$ (10)

The Hamiltonian is additive; hence, the joint distribution factorizes into a "potential energy" and a "kinetic energy":

$$p(X,\mu) = \exp\{-H(X,\mu)\} + \exp\left\{\frac{-\mu^T \mu}{2}\right\}\exp\{-U(X)\}$$ (11)

To sample the conformations X from their conditional posterior, $p(X|\cdot)$, we simply draw samples (X,μ) from $p(X,\mu)$ and discard the momentum variables. The HMC scheme simulates the joint distribution (11) as follows:

1. Draw generalized momenta from a unit-variance normal distribution (equivalent to assigning starting velocities in MD calculations);
2. Calculate an MD trajectory starting at X and μ to generate a proposal coordinates and momenta X' and μ';
3. Apply the standard Metropolis criterion on the total energy H, that is, accept X' and μ' with probability $P = \min\{1, \exp(-\Delta H)\}$, where $\Delta H = H(X',\mu') - H(X,\mu)$.

The total energy, H, is conserved under Hamiltonian dynamics; this leads to high acceptance probabilities. The algorithm exploits small numerical errors in the MD trajectory to generate states with $\Delta H \neq 0$.

To numerically integrate the Hamiltonian equations, we use the standard leapfrog discretization scheme directly in torsion angle space, i.e. we work with a diagonal instead of a non-diagonal mass matrix arising from the transformation of Cartesian coordinates to torsion angles. The diagonal mass matrix leads to an unphysical trajectory in Cartesian space (angular momentum is not conserved). This is not a problem for our application since the MD trajectory is only used to generate proposal states for the Metropolis criterion; the trajectory itself is discarded.

3.   Applications

3.1.   STRUCTURE DETERMINATION WITH A SPARSE NOE DATA SET

We applied the ISD approach to NMR measurements on the Fyn SH3 domain (59 amino acids, 275 dihedral angles). Distances between amide protons were estimated from NOE peak intensities measured on {15N, 2H} enriched protein [15]. The data comprises 48 sequential, 36

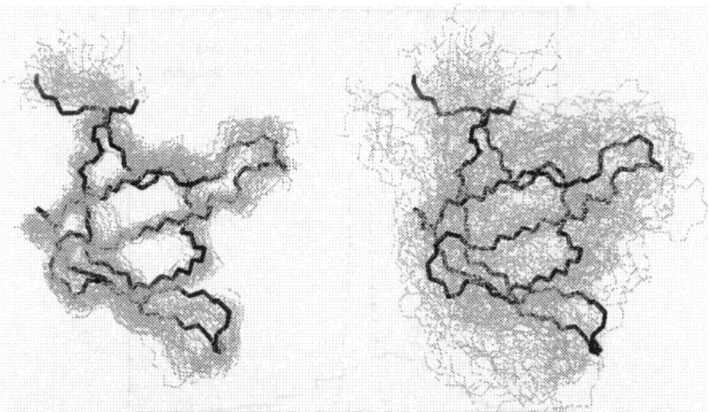

Figure 10. Structure ensembles. Backbone traces of conformational samples obtained by MCMC sampling (left hand side) and a structure ensemble generated by repeatedly running a standard simulated annealing protocol [16, 17] for NMR structure calculation with fixed weight w_{data} (right hand side) are shown in *grey*. The X-ray structure of the SH3 domain is drawn as a thick *black line*. Superposition and plotting of the structures was carried out with MOLMOL [18].

short-range, 4 medium- and 66 long-range NOEs; 154 in total. This is a sparse, poorly conditioned data set, since on average only one measured long-range distance is available for each amino acid.

Compared to conventional structure ensembles, the posterior ensemble obtained by Markov chain Monte Carlo (MCMC) sampling is better defined and systematically closer to the X-ray structure [19] (Figure 10): A comparison of the backbone heavy atoms N, C_a, and C yields an RMSD (root mean square deviation) value of 1.54 ± 0.14 Å and an RMSD of 1.10 ± 0.13 Å for the secondary structural elements, respectively. Our result compares well to the structures calculated by Mal et al. [15] using the same data: They determined an ensemble of structures with an overall RMSD of 2.86 ± 0.33 Å and an RMSD of 2.01 ± 0.28 Å for secondary structure elements.

The MCMC algorithm estimates the unknown scale γ and error σ along with the structure (Figure 11). As the estimation procedure is consistent, stable posterior histograms are obtained. In conventional approaches, the hybrid energy is an objective function for the coordinates only. Thus, in order to determine nuisance parameters quantitatively, meta-algorithms must be devised. Cross-validation [20, 21], for example, can be used to set the weight $w_{data} \propto \sigma^{-2}$. Cross-validation is often also used to determine the quality of a structure. The Bayesian procedure derives statistically meaningful error bars for all hypothesis parameters, especially the atom positions (Figure 12).

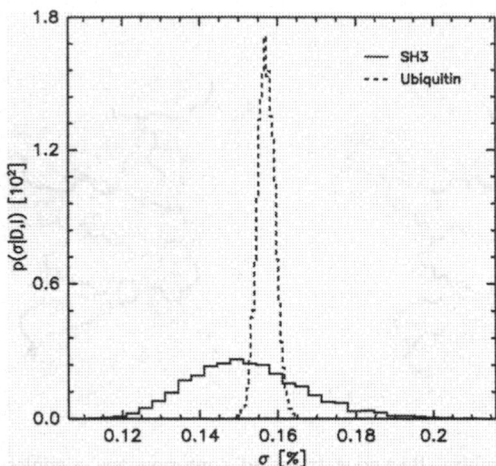

Figure 11. Estimation of auxiliary quantities. Marginal posterior $p(\sigma|D,I)$ distribution for the nuisance parameter σ (error of the log-normal model) obtained from Monte Carlo samples. In traditional approaches, this parameter would have to be determined by cross-validation.

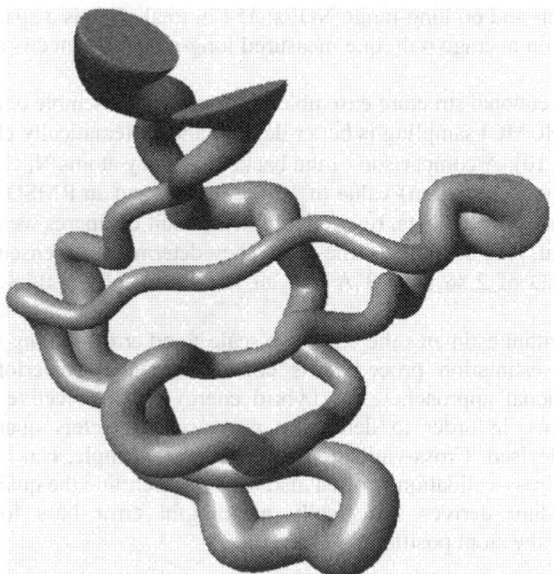

Figure 12. Conformational uncertainty. MOLMOL "sausage" plot of the mean structure with atom-wise error bars indicated by the thickness of the sausage. The atom positions of both the termini (*top*) and the loop regions (*bottom* and right hand side) show significant uncertainties.

3.2. DETERMINATION OF ϕ ANGLES FROM COUPLING CONSTANTS

The Bayesian formulation is a general framework not restricted to full determination of structures. For example, the problem of unknown Karplus coefficients can be treated quite elegantly: The observation of a scalar coupling constant of strength ^3J is described by a probability which expresses the fact that measured and theoretically predicted scalar couplings will never match exactly, due to experimental and processing errors as well as theoretical shortcomings [22]. As already mentioned in Eq. (3), the Gaussian is the appropriate distribution to describe this discrepancy.

This probability density is conditioned on the actual torsion angle, on the parameters of the Karplus curve and on the global error σ. If we observe n couplings $D = \{^3J_1, ..., ^3J_n\}$ independently, the likelihood function, i.e. the probability of all measurements, will be

$$L(\varphi, A, B, C, \sigma) = (2\pi\sigma^2)^{-n/2} \exp\{-\frac{1}{2\sigma^2} \chi^2(\varphi, A, B, C)\}$$ (12)

where the misfit between data and hypothesis parameters is

$$\chi^2(\varphi, A, B, C) = \sum_{i=1}^{n} \left(^3J_i - A\cos^2\varphi_i - B\cos\varphi_i - C \right)^2.$$ (13)

If we have only the n scalar coupling measurements, the parameters φ_i, A, B, C, and σ are unknown but can be estimated from the data. Any quantity that is necessary to calculate the observable, in this case, a coupling constant, can be reconstructed from the measurements. Not only the torsion angles but also the Karplus coefficients influence the result of the backcalculation of scalar coupling constants; the error quantifies the closeness of the fit. The calculation of the parameters from the data is possible by application of Bayes' theorem [6]. The likelihood determines the unknown parameters in terms of a posterior density, multiplied with a prior density $\pi(\varphi, A, B, C, \sigma)$,

$$p(\varphi, A, B, C, \sigma) \propto L(\varphi, A, B, C, \sigma)\pi(\varphi, A, B, C, \sigma).$$ (14)

From the posterior density we can not only derive the most probable torsion angles but also the Karplus coefficients A, B, C and the unknown error σ. Figure 13 shows the distributions of three representative torsion angles calculated from the data for ubiquitin [23]. Note that this distribution contains the dependency of a variation of the Karplus parameters A, B, C, and σ. Figure 14 shows the distributions of the parameters A, B, and C themselves.

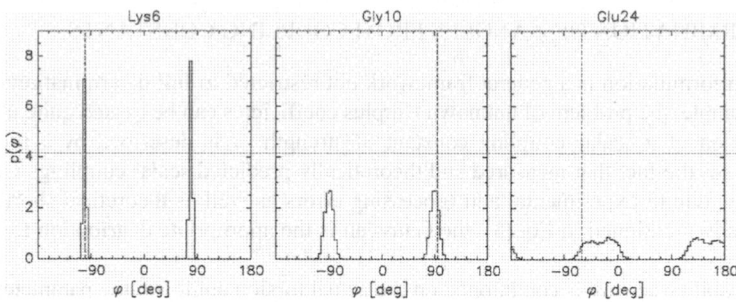

Figure 13. Posterior histograms for three representative φ angles. The values found in the NMR structure 1D3Z are indicated by *dotted lines*.

4.   Conclusions and outlook

A rigorous probabilistic approach to structure determination has decisive advantages. By considering the joint posterior of all unknown quantities, additional parameters are estimated along with the atomic coordinates. The method can be extended to more than the few nuisance parameters shown in this paper. For example, one can determine the data quality of a large number of data sets and other nuisance parameters (Karplus coefficients, Tensor parameters for residual dipolar couplings).

The method has no free parameter, hence, tedious and time-consuming searches for optimal values are no longer necessary. Once the model to describe the data has been chosen, the rules of probability theory uniquely determine the posterior distribution. It is then only a computational issue to generate posterior samples. Further user intervention is not necessary, and structure determination becomes more objective. The method is unbiased in the sense that if different people use the same data and the same model they get the same probability distribution.

While there are no free parameters to choose or optimize, there are a number of parameters that are important for the efficiency of the calculation. In particular one has to choose the set of replica parameters λ and q. These need to be chosen in such a way that the neighbouring trajectories have a sufficiently high probability to interchange. Since this probability depends on the overlap of neighbouring distributions and thus on the width of the posterior distribution, an optimal scheme will depend on both the system size and the size of the data set.

At first sight, ISD seems more time-consuming than minimization based techniques. While this is certainly true in terms of total CPU time (the replica-exchange algorithm typically requires a PC cluster and 3–5 days of computation) differences are less obvious on a per structure basis: once the MCMC algorithm has converged, the time required to calculate a single conformer is comparable to that needed by conventional methods (a matter of seconds). Furthermore, ISD gives richer results: neither optimal estimates of nuisance parameters including their uncertainty nor the uncertainty of the atomic coordinates are accessible with standard techniques.

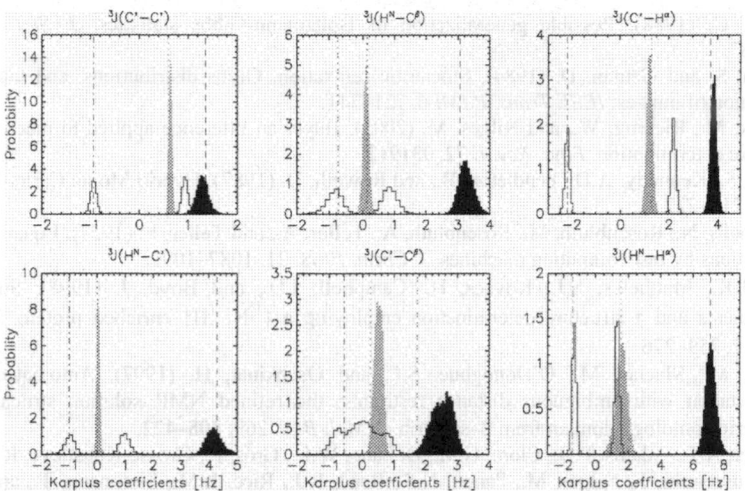

Figure 14. Posterior histograms for the Karplus coefficients of the six scalar couplings (*black*: *A*, *white*: *B*, *grey*: *C*). The estimates obtained by maximum likelihood using the NMR structure 1D3Z are indicated as *dashed lines*.

References

1. Karplus, M. (1963). Vicinal proton coupling in nuclear magnetic resonance. *J. Am. Chem. Soc.* 85, 2870–2871.
2. Pardi, A., Billeter, M., and Wüthrich, K. (1984). Calibration of the angular dependence of the amide proton-C_α proton coupling constants, $^3J_{HN\alpha}$, in a globular protein. *J. Mol. Biol.* 180, 741–751.
3. Güntert, P., Braun, W., Billeter, M., and Wüthrich, K. (1989). Automated stereospecific ^1H NMR assignments and their impact on the precision of protein structure determination in solution. *J. Am. Chem. Soc.* 111, 3997–4004.
4. Cox, R.T. (1946). Probability, frequency and reasonable expectation. *Am. J. Physics* 14(1), 1–13.
5. Rieping, W., Habeck, M., and Nilges, M. (2005). Structure determination from heterogeneous NMR data. In R. Fischer, R. Preuss, and U. von Toussaint (eds.), *Bayesian Inference and Maximum Entropy Methods in Science and Engineering*, American Institute of Physics, Springer, Heidelberg.
6. Jaynes, E.T. (2003). *Probability Theory: The Logic of Science*. Cambridge University Press, Cambridge, UK.
7. Rieping, W., Habeck, M., and Nilges, M. (2005). Inferential structure determination. *Science* 309, 303–306.
8. Jeffreys, H. (1946). An invariant form for the prior probability in estimation problems. *Proc. Roy. Soc.* A186, 453.
9. Habeck, M., Rieping, W., and Nilges, M. (2005). Estimation of proton configurations from NOESY spectra. In R. Fischer, R. Preuss, and U. von Toussaint (eds.), *Bayesian Inference and Maximum Entropy Methods in Science and Engineering*, Springer, Heidelberg.

10. Tsallis, C. (1988). Possible generalization of Boltzmann-Gibbs statistics. *J. Stat. Phys.* 52, 479–487.
11. Geman, S. and Geman, D. (1984). Stochastic relaxation, Gibbs distributions, and the Bayesian restoration of images. *IEEE Trans. PAMI* 6, 721–741.
12. Habeck, M., Rieping, W., and Nilges, M. (2005). Bayesian inference applied to macromolecular structure determination. *Phys. Rev. E* 72, 031912.
13. Duane, S., Kennedy, A.D., Pendleton, B., and Roweth, D. (1987). Hybrid Monte Carlo. *Phys. Lett. B* 195, 216–222.
14. Metropolis, N., Rosenbluth, M., Rosenbluth, A., Teller, A., and Teller, E. (1957). Equation of state calculations by fast computing machines. *J. Chem. Phys.* 21, 1087–1092.
15. Mal, T.K., Matthews, S.J., Kovacs, H., Campbell, I.D., and Boyd, J. (1998). Some NMR experiments and a structure determination employing a $\{^{15}N, ^2H\}$ enriched protein. *J. Biomol. NMR* 12, 259–276.
16. Nilges, M., Macias, M., O'Donoghue, S.I., and Oschkinat, H. (1997). Automated NOESY interpretation with ambiguous distance restraints: the refined NMR solution structure of the pleckstrin homology domain from β–spectrin. *J. Mol. Biol.* 269, 408–422.
17. Brunger, A.T., Adams, P.D., Clore, G.M., DeLano, W.L., Gros, P., Grosse-Kunstleve, R.W., Jiang, J.S., Kuszewski, J., Nilges, M., Pannu, N.S., Read, R.J., Rice, L.M., Simonson, T., and Warren, G.L. (1998). Crystallography and NMR system: a new software suite for macromolecular structure determination. *Acta Crystallogr. D* D54, 905–921.
18. Koradi, R., Billeter, M., and Wüthrich, K. (1996). MOLMOL: a program for display and analysis of macromolecular structures. *J. Mol. Graphics* 14, 51–55.
19. Noble, M., Musacchio, A., Saraste, M., and Wierenga, R. (1993). Crystal structure of the SH3 domain in human Fyn; comparison of the three-dimensional structures of SH3 domains in tyrosine kinases and spectrin. *EMBO J.* 12, 2617–2624.
20. Brunger, A.T. (1992). *X–plor. A System for X-ray Crystallography and NMR.* Yale University Press, New Haven, CT.
21. Brunger, A.T., Clore, G.M., Gronenborn, A.M., Saffrich, R., and Nilges, M. (1993). Assessing the quality of solution nuclear magnetic resonance structures by complete cross-validation. *Science* 261, 328–331.
22. Habeck, M., Rieping, W., and Nilges, M. (2005). Bayesian estimation of Karplus parameters and torsion angles from three-bond scalar couplings constants. *J. Magn. Reson.* 177, 160–165.
23. Cornilescu, G., Marquardt, J.L., Ottiger, M., and Bax, A. (1998). Validation of protein structure from anisotropic carabonyl chemical shifts in a dilute liquid crystalline phase. *J. Am. Chem. Soc.* 120, 6836–6837.

MOLECULAR INSIGHTS INTO PKR ACTIVATION BY VIRAL DOUBLE-STRANDED RNA

SEAN A. MCKENNA[1], DARRIN A. LINDHOUT[1], COLIN ECHEVERRÍA AITKEN[2], AND JOSEPH D. PUGLISI[1,3]
Department of Structural Biology[1]
Biophysics Program[2]
Stanford Magnetic Resonance Laboratory[3],
Stanford University School of Medicine,
Stanford, CA94305-5126, USA

1. Introduction

The presence of highly structured RNAs within viral genomes is a key extracellular and intracellular indicator of infection. As a result, double-stranded RNA-dependent protein kinase (PKR) modulates host-mediated down-regulation of eukaryotic translation.[1] Upon binding double-stranded RNA molecules (dsRNA), PKR can become activated and phosphorylate cellular targets, including translation initiation factor 2α (eIF-2α) at Ser51.[2] Phosphorylation at this site inhibits the activity of guanine nucleotide exchange factor eIF2B. As such, the pool of active eIF2 ternary complex decreases, causing a general decrease in translation initiation.[3] PKR has also been implicated in cell growth,[4] signal transduction,[5] induction of apoptosis,[6] and RNA interference.[7]

Human PKR is a 551-residue enzyme consisting of 3 distinct regions (Figure 1). The 20 kDa RNA binding domain from PKR (dsRBD1/2) contains two highly similar dsRNA binding domains responsible for dsRNA recognition.[8] High-affinity PKR-dsRNA interaction requires both dsRBDs of PKR, as well as specific structural and length requirements of the dsRNA.[9–11] The C-terminal region of PKR is a serine-threonine kinase domain that is responsible for phosphorylation activity and substrate recognition.[12] An 80-residue interdomain linker connects the N- and C-terminal domains of PKR. Latent PKR undergoes *trans*-autophosphorylation upon dsRNA binding, which greatly increases kinase activity.[13–14] Dimerization of PKR may be required for efficient kinase activation,[15–17] as both dsRNA binding and phosphorylation state of the protein may affect dimerization equilibrium.[18] Despite these data,

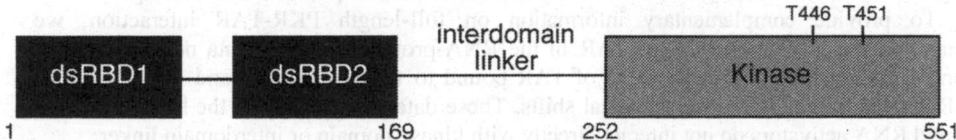

Figure 1. **Domain organization of PKR.** N-terminal dsRBDs, C-terminal kinase domain, and the interdomain linker are shown. Critical autophosphorylation sites (T446, T451) in the kinase domain are indicated.

J. D. Puglisi (ed.), Structure and Biophysics – New Technologies for Current Challenges in Biology and Beyond, 99–110.

the mechanism of PKR kinase activation remains unresolved. Here we perform experiments to provide a molecular basis for the activation of PKR. The results suggest a mechanism of PKR activation in which RNA binding and autophosphorylation modulate protein self-association and activity.

2.   PKR Displays an Open Conformation

To explore the interplay between the RNA-binding and kinase domains, we performed NMR studies on full-length PKR. The ^{15}N-TROSYHSQC spectra of latent full-length ^{2}H^{15}N-PKR revealed a sufficiently resolved spectrum, which allowed spectral identification of a large number of amide resonances (Figure 2A). The near-perfect spectral superimposition of free full-length PKR with either dsRBD1/2, or the kinase domain resonances (data not shown) suggests that the individual domains in the context of full-length PKR are behaving independently in solution. These data suggest that the RNA binding and kinase domains do not significantly interact in latent PKR.

Spectral overlay allowed assignment of 137 out of 169 resonances in the RNA-binding domains, and 82 out of 299 kinase domain resonances, predominantly in the outlying regions of the spectra.[19] The 80-residue linker region between the dsRBDs and kinase domain is unassigned as it is unstructured in solution, present with the bulk signal in the overlapping region of the spectra.

The interaction between dsRNA activator (HIV-TAR, Figure 3) and full-length PKR was characterized using NMR. Chemical shift perturbations to full-length PKR upon formation of a 1:1 complex with HIV-TAR occur predominantly within the RNA binding domains (Figure. 2A). Significant perturbations occur to resonances within both dsRBD domains (i.e. L25, L30, K136) with minimal perturbations to kinase resonances (i.e. K261, Q365, I503) (Figure 2B). In dsRBD1, two surface-exposed regions appear important: the first consisting of residues from helix α1 and the loop that precedes this helix, and a second containing residues from the loop connecting ß1 to ß2 and the region N-terminal to helix α2. In dsRBD2, a single surface exposed region is perturbed, consisting of residues from helix α1 and the loop connecting it to ß1. The linker connecting the dsRBDs also demonstrates significant perturbations. Resonances within the dsRBD domains closely match those previously reported for a dsRBD1/2•TAR complex,[9] indicating that the binding of dsRBD1/2 to TAR in the presence or absence of the kinase domain are similar. The PKR kinase resonances display near-perfect overlap when compared to the unbound protein. These results show that the kinase domain does not directly participate in the binding of dsRNA, and suggest that no large conformational change within the kinase domain occurs upon activator binding.

To provide complementary information on full-length PKR-TAR interaction, we examined the NMR spectrum of TAR in the RNA-protein complex (data not shown). A comparison of the 1D-imino spectra of TAR bound to full-length PKR and TAR bound to dsRBD1/2 show very similar chemical shifts. These data further support the hypothesis that bound RNA activators do not interact directly with kinase domain or interdomain linker.

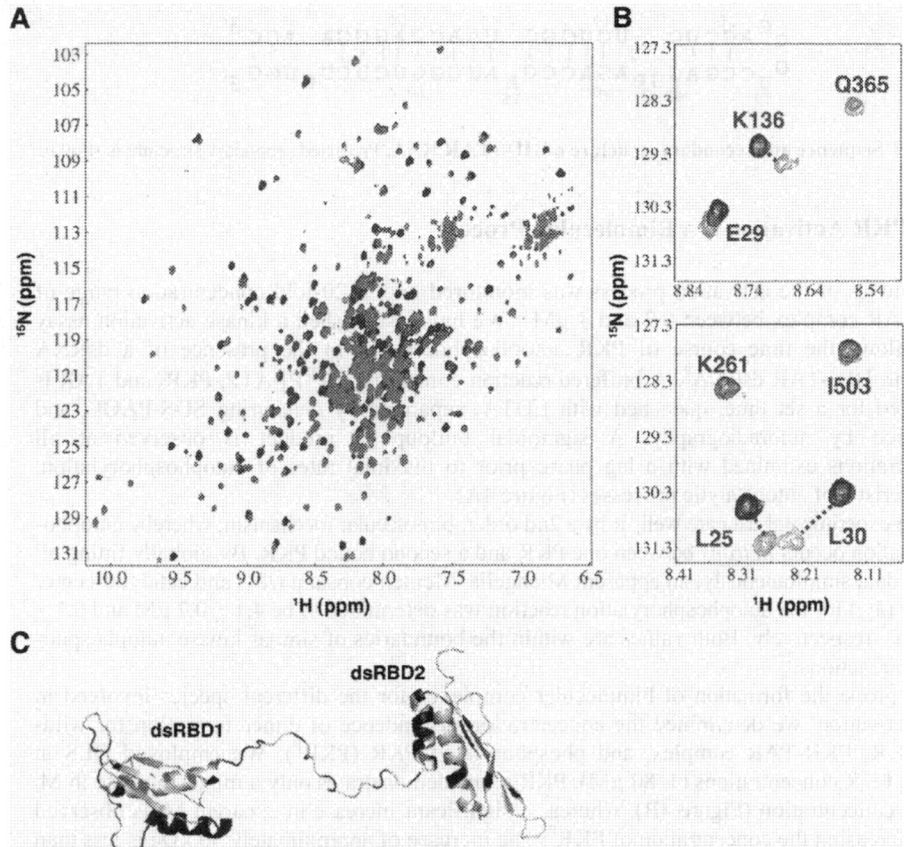

Figure 2. **NMR of PKR in the presence and absence of TAR RNA.** (A) ^{15}N-TROSYHSQC spectral overlay of unbound PKR1-551 (*black*) and TAR bound PKR1-551 (*grey*). NMR spectral data was obtained using a Varian Unity INOVA 800MHz spectrometer at 30°C, and acquired using the sensitivity-enhanced gradient pulse scheme in Biopack (Varian Inc.). (B) Detailed view of the spectra shown in (A), demonstrating chemical-shift perturbations only to the RNA binding resonances within PKR1-551 upon TAR binding. (C) Weighted average changes in ^{1}H and ^{15}N chemical-shift changes were determined for each residue relative to free dsRBD1/2. Amino acid residues affected by RNA binding are shown in *black* on the 3D structure of PKR1-169. The total average change in backbone amide ^{1}H and ^{15}N chemical shift was determined according to the following equation:

$$\Delta\partial_{\text{total}} = \sqrt{\left(0.1\Delta\partial^{15}N\right)^{2} + \left(\Delta\partial^{1}H\right)^{2}}.$$

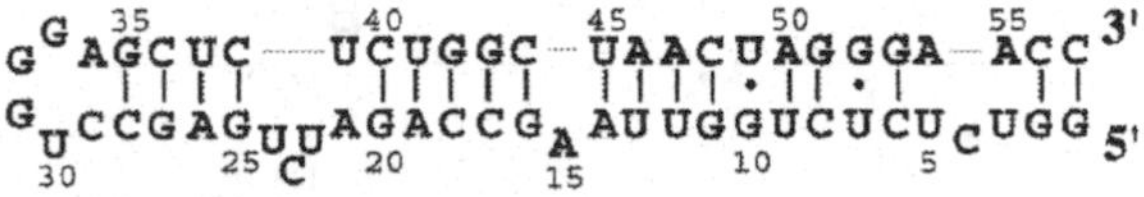

```
   G   35          40           45          50          55
  G  AGCUC ---- UCUGGC --- UAACUAGGGA --ACC  3'
      | | | |       | | | | |    | | | | • | | • |      | |
  G                AGACCG   AU UGGUCUCU    UGG  5'
   U CCGAG U U         A                C
    30    25 C   20         15         10      5
```

Figure 3. **Sequence and secondary structure of HIV-TAR RNA.** Predicted secondary structure is shown.

3.   PKR Activation is a Bimolecular Process

The kinetics of the activation process was monitored over a 20-fold concentration range of PKR-TAR complex between 0.2 and 4 µM. We have established a kinase activation assay that follows the time course of PKR autophosphorylation in the presence of a dsRNA activator, HIV-TAR dsRNA[9]. A buffered reaction containing $[\gamma\text{-}^{32}P]$-ATP, PKR, and TAR is incubated for a set time, quenched with EDTA, separated by denaturing SDS-PAGE, and quantified by autoradiography. A sigmoidal buildup of product is observed at all concentrations examined with a lag phase prior to maximal rates of autophosphorylation, characteristic of autocatalytic processes (Figure 4A).

The experimental data are well fit by a 2nd order, bimolecular mechanism, whereby autophosphorylation occurs in *trans* between one PKR and a second bound PKR. By globally fitting all kinetic data simultaneously, an apparent Michaelis–Menten constant (K_M) and catalytic-center activity (k_{cat}) for the autophosphorylation reaction was determined to be 4.1 ± 0.7 µM and 0.7 ± 0.3 min^{-1} respectively. Both values are within the boundaries of similar kinase autophosphorylation reactions.[20]

To probe the formation of bimolecular complexes for the different species involved in PKR activation, we determined the concentration dependence of dimer formation for wild-type PKR, PKR-TAR complex, and phosphorylated PKR (PKR^P). We employed DLS at various PKR concentrations (1–80 µM). PKR alone demonstrates only a minor increase in M_r at high concentration (Figure 4B), whereas a significant increase in apparent M_r is observed upon increasing the concentration of PKR^P. The increase of approximately 30 kDa is less than would be expected for a dimer of two PKR molecules (136 kDa). PKR dimerization may yield in a nonspherical complex, or, more likely, the values observed may reflect an equilibrium between monomer and dimer. A 1:1 PKR^P-PKR complex gave a similar increase in M_r when compared to PKR^P alone, indicating a similar affinity for phosphorylated and dephosphorylated enzyme. A 1:1 PKR-TAR complex gave an intermediate increase in M_r when compared to PKR or PKR^P, indicating that dsRNA-bound PKR has a greater self-association constant than unbound PKR, but weaker than phosphorylated PKR. The hydrodynamic behavior of PKR and complexes determined by DLS was confirmed using size-exclusion chromatography.

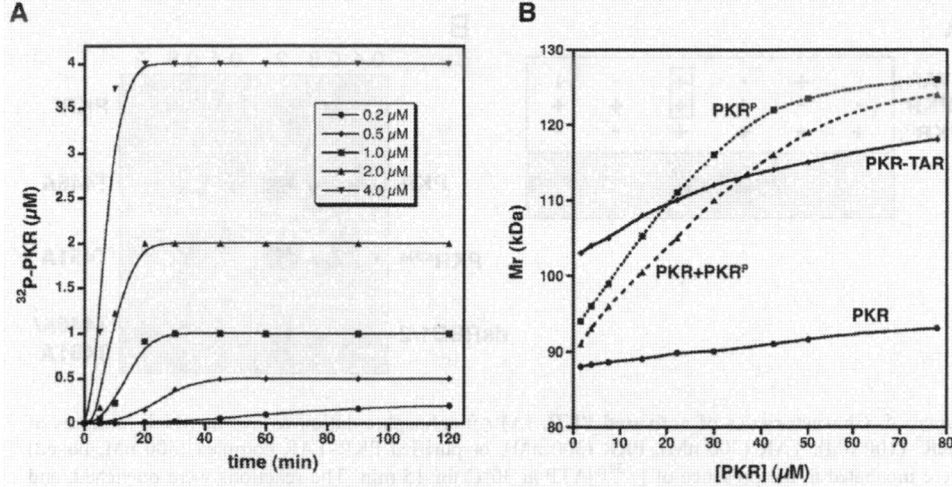

Figure 4. **Activated PKR forms a weak dimer.** (A) Progress curves of PKR autophosphorylation in the presence of increasing concentrations of PKR-TAR complex. Each time point was performed in triplicate, resolved by SDS-PAGE, and quantified for ^{32}P incorporation by autoradiography. Data was analyzed using Berkeley Madonna X (version 8.3.12) software to fit the differential equations that describe the bimolecular model of PKR activation. We solved the differential equations for $E^*(t)$, the rate of formation of phosphorylated PKR product, using the Runge–Kutta algorithm. (B) Concentration-dependent dimerization of PKR was examined by determining the molecular weight at the specified concentration of PKR (*circles*), PKR^P (*squares*), PKR-TAR (*diamonds*), or equimolar mixture of PKR^P and PKR (*triangles*). Experiments were performed at 30°C with a DynaPro-801 molecular sizing instrument (Protein Solutions Co.). For both (A) and (B), each data point was repeated in triplicate. Error bars are not shown for clarity, but errors were typically less than 5% of the value shown.

4.   Characterization of the Phosphorylated State of PKR

To examine the properties of phosphorylated PKR, we produced phosphorylated PKR (PKR^P) in an RNA-independent fashion. PKR at high concentration (>15 μM) was incubated in the presence of ATP and Mg^{2+}, and subsequently phosphorylated PKR is separated from ATP and Mg^{2+} using size exclusion chromatography. Wild-type PKR and PKR^P were used both as substrates for autophosphorylation and as kinases capable of *trans*-phosphorylating wild-type PKR (Figure 5A). After 15-min incubation, minimal phosphorylation of PKR^P is observed when incubated in the presence of TAR activator, indicating that PKR^P is unresponsive to dsRNA, and does not further phosphorylate itself under these conditions. PKR^P efficiently *trans*-phosphorylates wild-type latent PKR as a substrate. A 1:1 PKR-TAR complex is also an efficient substrate for *trans*-autophosphorylation by PKR^P, indicating that dsRNA binding to PKR does not block phosphorylation. *Trans*-autophosphorylation of PKR by PKR^P occurs at a 20-fold faster rate than dsRNA-mediated autophosphorylation. Thus, purified PKR^P serves as an extremely potent activator of latent PKR.

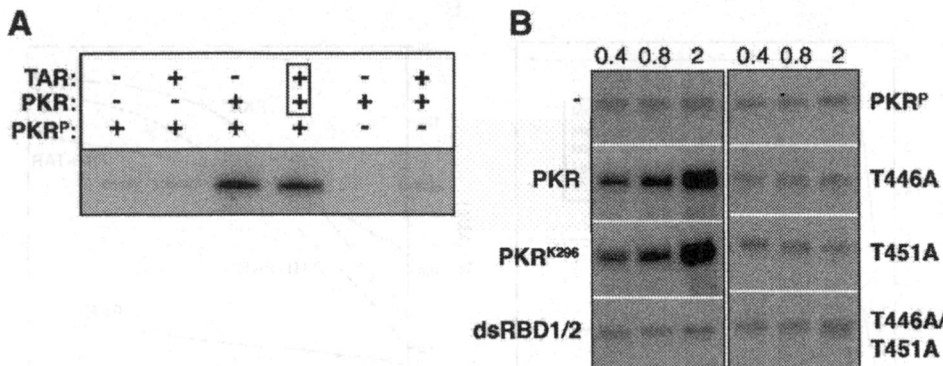

Figure 5. **Characteristics of activated PKR.** (A) Autophosphorylation assays in which mixtures of PKR^P (100 nM), TAR (300 nM), PKR (300 nM), or purified PKR-TAR complex (300 nM, boxed) were incubated in the presence of [γ-^{32}P]ATP at 30°C for 15 min. The reactions were quenched, and resolved by SDS-PAGE. Gels were quantified by autoradiography to quantify the extent of PKR autophosphorylation. (B) *Trans*-autophosphorylation assays in which PKR^P (100 nM) was added to a reaction including [γ -^{32}P]-ATP and the specified PKR derivative at concentrations of 0.4, 0.8, and 2 µM respectively for 10 min at 30°C.

PKR mutations affect substrate activity for *trans* autophosphorylation by PKR^P (Figure 5B). As expected, wild-type PKR serves as an efficient substrate, whereas PKR^P does not. A PKR derivative containing a catalytically inactive mutation in the ATP binding site, PKR^{K296R}, is also phosphorylated efficiently by PKR^P, indicating that a functional ATP binding and active site is not required in a substrate for phosphorylation. Conversely, mutation at either position T446 or T451, the critical phosphorylation sites required for activation, results in attenuation of *trans*-autophosphorylation. These results suggest potential cooperativity between T446 and T451, a plausible hypothesis based on their structural proximity.[12] Isolated dsRBDs alone are not *trans*-autophosphorylated, indicating that they are not targets for initial activation.

5. Structural Features of the Phosphorylated State

The structural properties of phosphorylated PKR were next investigated using NMR spectroscopy. The ^{15}N-TROSYHSQC of PKR^P reveals a homogeneous sample, indicative of a single phosphorylation state of the protein (Figure 6A); doubling of resonances was not observed. Broader resonances were observed when compared to latent PKR, suggesting an increased rotational correlation time, indicative of a monomer/dimer equilibrium in solution. Resonance assignments were performed by overlay with the latent protein and assuming nearest-neighbor shifts. As with the unbound protein, there were no assignments for the linker region of the protein (170–251). dsRBD1/2 resonances (1–169) undergo minimal perturbation upon PKR activation, indicating that they are not phosphorylated in the activated form or in contact with the kinase domain.

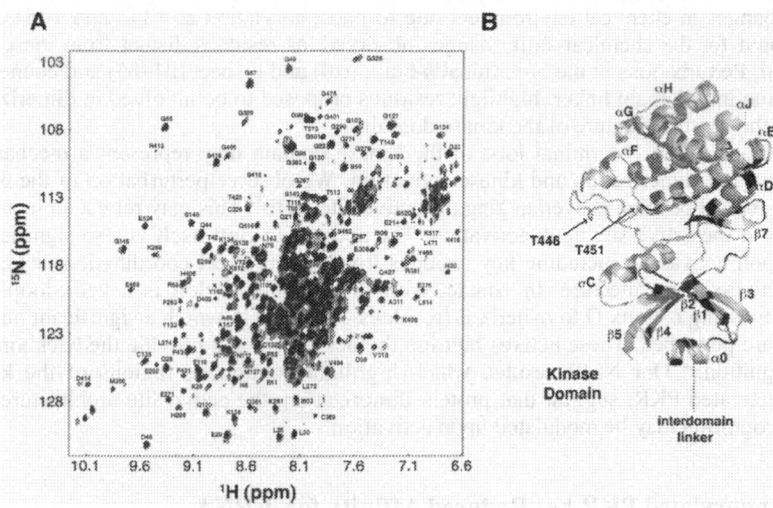

Figure 6. **NMR of phosphorylated PKR.** (A) ^{15}N-TROSY HSQC spectral overlay of PKR in latent (*black*) and activated (*grey*) form. (B) $\Delta\delta_{total}$ values were determined for each residue of PKR[P] relative to latent full-length PKR, and amino acids affected by activation are colored black on the 3D structure of the kinase domain.[12]

Numerous resonances from the kinase domain are perturbed in activated PKR (Figure 6B). Backbone amide groups that exhibit the largest perturbations cluster to three contiguous surface exposed regions of the protein: around the N-terminal ß-sheet, the kinase active site, and the region proximal to the eIF2α substrate-binding site. Residues from the N-terminal lobe of PKR undergo significant chemical-shift perturbations, indicating a potential conformational change to the N-terminal ß-sheet upon phosphorylation. These include residues from ß1 (L272, G274), ß1–ß2 loop (S275, G276, G277, G279), ß2 (F282, K285), ß2–ß3 loop (H286, R287), ß3 (I294, V295), ß4 (N324, G325, C326, G329), and ß5 (Q365). Interestingly, the N-terminal α-helix of the kinase domain, which is directly connected to the interdomain linker, is also perturbed (K261, F263). The region surrounding the ATP-binding site, centered around K296, is also significantly perturbed, and includes residues from ß6 to ß7 loop (R413, D414, K416, S418), ß7 (I420, F421), ß7–ß8 loop (T425), ß8 (Q427, I430) and the loop C-terminal to ß8 (D432). Several of these residues have been directly implicated in ATP coordination, including R413, D414, and D432.[12] A two-helix stretch which flanks the proposed eIF2a binding site[12] is also perturbed, and encompasses aD (E375, R381), αD-αE loop (G383), aE (L390, Q397, G401, D403, H406, K408), and αF-αG loop (V484).

Local changes in chemical environment due to phosphorylation at T446 and T451 alone cannot account for the chemical-shift changes observed, as residues distant from these sites are perturbed. Perturbations in the N-terminal a-helix (α0) and ß-sheet (ß1–ß5) and connecting loops, near the interdomain linker, highlight residues proposed to be involved in dimerization observed in the crystal structure of the kinase domain.[12]

Perturbations to the N-terminal lobe of the kinase domain may represent a mechanistic link between PKR dimerization and kinase activation. We observe perturbation of the ß1–ß2 loop, whose typical position is overhanging the active-site cleft.[12] This may reflect an increased accessibility to the active site upon activation. The active-site cleft itself is also significantly perturbed upon activation, including key residues involved in direct coordination or maintenance of active-site architecture. In contact with the active-site cleft is a helix-loop-helix motif encompassing α-helix D to α-helix E (residues 375–410) in which a significant number of residues are perturbed. These helixes buttress the active-site and provide the backbone for eIF2α recognition.[12] Our NMR results, which highlight contiguous regions of the kinase domain in activated PKR, suggest that protein dimerization, the active-site architecture, and substrate recognition may be modulated upon activation.

6. Phosphorylated PKR has Reduced Affinity for dsRNA

The communication between dsRNA binding and PKR activation was further explored by monitoring the stability of PKR-dsRNA complex during activation. To observe RNA release directly, native gel shift mobility assays were performed on HIV-TAR dsRNA in the presence of PKR under activating (ATP/MgCl$_2$) and nonactivating (AMPPNP/MgCl$_2$) conditions (Figure 7A). Upon incubation at temperatures sufficient for PKR activation (90 min), disassociation of the PKR-TAR complex is observed when both ATP and MgCl$_2$ are present. When a nonhydrolysable ATP analogue (AMPPNP) is employed, no RNA release is observed. RNA release closely parallels activation, as almost complete dissociation is observed at the midpoint of the sigmoidal activation progress curve at all concentrations examined (Figure 7B); RNA activators dissociate rapidly from PKR, with rates >1 s^{-1} (data not shown).

Native gel shift mobility assays were used to quantify dsRNA release from both wild-type and mutant PKR. In all cases where AMPPNP was used, only minor RNA release is observed (Figure 7C). However, under activating conditions, only wild-type PKR releases RNA efficiently, whereas ATP binding site mutants (PKR^{K296R}) and activation loop phosphorylation site mutants (PKR^{T446A} and PKR^{T451A}) do not. Therefore, dissociation of dsRNA activator from PKR is coincident with activation. The molecular basis for RNA release remains unclear, but demonstrates thermodynamic and mechanistic coupling between kinase and RNA binding domains.

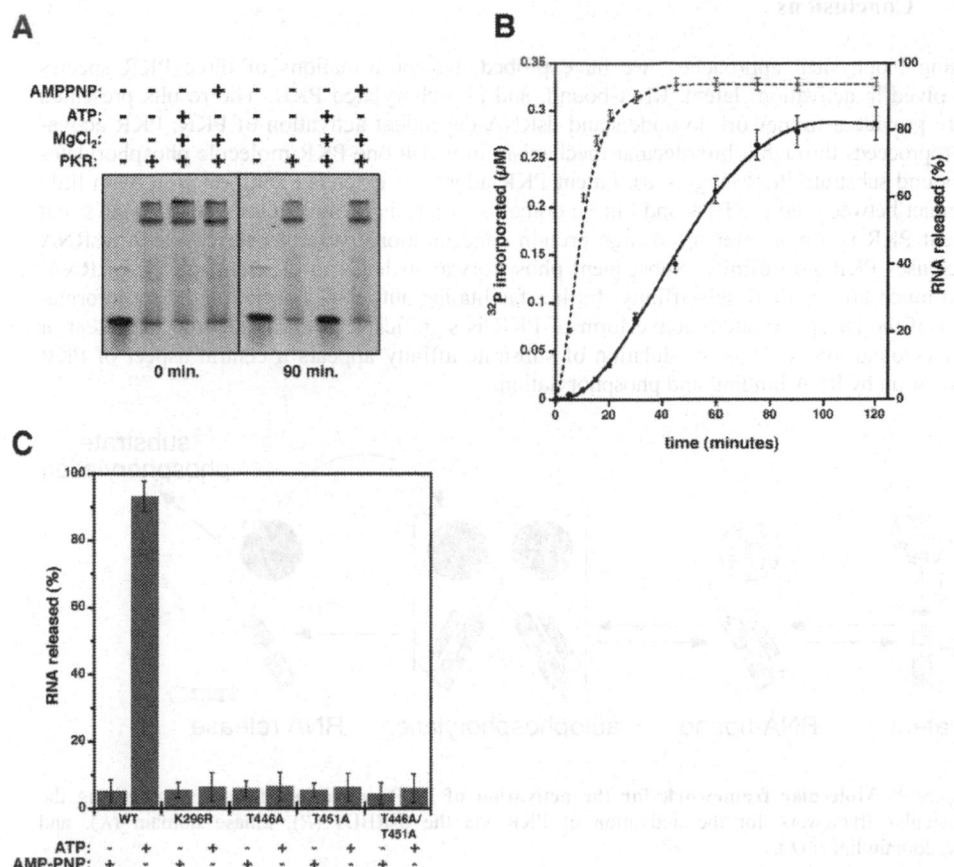

Figure 7. **Activator RNA is released upon PKR activation.** (A) Native gel mobility-shift for TAR (200 nM) binding to PKR (300 nM) in reactions containing ATP (1 mM), AMPPNP (1 mM), and MgCl₂ (2 mM). Samples were incubated at 30°C for the time indicated, resolved on 5% TBE gel, and visualized by SybrGreenII staining. (B) PKR (200 nM) was incubated [γ-³²P] ATP in the presence of TAR (300 nM), and aliquots were removed an resolved by either SDS-PAGE and quantified for autophosphorylation (*left axis, solid black line*) or RNA release by native gel mobility-shift (*right axis, dashed line*) as in (A). (C) PKR-TAR complexes were pre-assembled (200 nM), and incubated at 30°C for 90 min in the presence or absence of ATP (1 mM) and MgCl₂ (1 mM). RNA release was quantified by resolving reaction components on native 5% TBE gels and dsRNA staining by SybrGreenII.

7.   Conclusions

Using biophysical approaches, we have probed the conformations of three PKR species involved in activation: latent, RNA-bound, and phosphorylated PKR. The results presented here provide a framework to understand dsRNA-dependent activation of PKR. PKR activation proceeds through a bimolecular mechanism in which one PKR molecule phosphorylates a bound substrate PKR (Figure 8). Latent PKR adopts an extended conformation, with little contact between the dsRBDs and kinase domains. Analysis of molecular weights shows that latent PKR is a monomer up to high protein concentrations, whereas activation by dsRNA increases PKR self-affinity. Subsequent phosphorylation leads to efficient release of RNA, and much greater PKR self-affinity, further facilitating autophosphorylation. The conformation of the phosphorylated, active form of PKR is significantly perturbed from the latent or RNA-bound forms. Thus, modulation of substrate affinity appears a central aspect of PKR activation by RNA binding and phosphorylation.

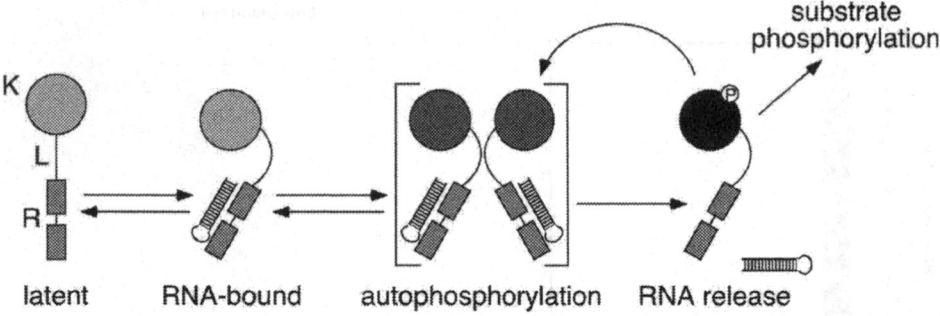

Figure 8. **Molecular framework for the activation of PKR activation.** Model summarizing the molecular framework for the activation of PKR via the dsRBDs (*R*), kinase domain (*K*), and interdomain linker (*L*).

Electrostatic changes may be the key to PKR activation. The surfaces of both dsRBDs and kinase domain near the active site are positively charged. The extended conformation of PKR may result from electrostatic repulsion between the two domains. Binding of RNA changes the electrostatic potential of the dsRBDs, and may alter the disposition of the linker to allow initial autophosphorylation. Covalent phosphorylation leads to RNA release, through reciprocal electrostatic signaling, and stabilizes an active form for PKR with high affinity for substrate. Highly charged molecules such as heparin also lead to activation of PKR,[21] consistent with a change in electrostatic potential modulating PKR self-association. More detailed structural studies, coupled with investigations of inter- and intra-domain dynamics, are needed to unravel the molecular details of PKR activation.

8. Acknowledgements

We thank M. Margaris for assistance, and other members of the Puglisi laboratory members for their help and advice. Supported by NIH AI47365 and GM078346. S.A. McKenna is supported by the Canadian Institutes of Health Research and the Alberta Heritage Foundation for Medical Research. D.A. Lindhout is supported by the Alberta Heritage Foundation for Medical Research.

References

1. Gale, M.J., Jr., Korth, M.J., and Katze, M.G. (1998). Repression of the PKR protein kinase by the hepatitis C virus NS5A protein: a potential mechanism of interferon resistance. *Clin. Diagn. Virol.* 10, 157–162.
2. Chong, K.L., Feng, L., Schappert, K., Meurs, E., Donahue, T.F., Friesen, J.D., Hovanessian, A.G., and Williams, B.R. (1992). Human p68 kinase exhibits growth suppression in yeast and homology to the translational regulator GCN2. *Embo. J.* 11, 1553–1562.
3. Dever, T.E. (2002). Gene-specific regulation by general translation factors. *Cell* 108, 545–556.
4. Srivastava, S.P., Davies, M.V., and Kaufman, R.J. (1995). Calcium depletion from the endoplasmic reticulum activates the double-stranded RNA-dependent protein kinase (PKR) to inhibit protein synthesis. *J. Biol. Chem.* 270, 16619–16624.
5. Mundschau, L.J. and Faller, D.V. (1995). Platelet-derived growth factor signal transduction through the interferon-inducible kinase PKR. Immediate early gene induction. *J. Biol. Chem.* 270, 3100–3106.
6. Der, S.D., Yang, Y.L., Weissmann, C., and Williams, B.R. (1997). A double-stranded RNA-activated protein kinase-dependent pathway mediating stress-induced apoptosis. *Proc. Natl. Acad. Sci. U S A* 94, 3279–3283.
7. Andersson, M.G., Haasnoot, P.C., Xu, N., Berenjian, S., Berkhout, B., and Akusjarvi, G. (2005). Suppression of RNA interference by adenovirus virus-associated RNA. *J. Virol.* 79, 9556–9565.
8. Nanduri, S., Carpick, B.W., Yang, Y., Williams, B.R., and Qin, J. (1998). Structure of the double-stranded RNA-binding domain of the protein kinase PKR reveals the molecular basis of its dsRNA-mediated activation. *Embo. J.* 17, 5458–5465.
9. Kim, I., Liu, C.W., and Puglisi, J.D. (2006). Specific recognition of HIV TAR RNA by the dsRNA binding domains (dsRBD1-dsRBD2) of PKR. *J. Mol. Biol.* 358, 430–442.
10. McCormack, S.J., Ortega, L.G., Doohan, J.P., and Samuel, C.E. (1994). Mechanism of interferon action motif I of the interferon-induced, RNA-dependent protein kinase (PKR) is sufficient to mediate RNA-binding activity. *Virol.* 198, 92–99.
11. Bevilacqua, P.C. and Cech, T.R. (1996). Minor-groove recognition of double-stranded RNA by the double-stranded RNA-binding domain from the RNA-activated protein kinase PKR. *Biochem.* 35, 9983–9994.
12. Dar, A.C., Dever, T.E., and Sicheri, F. (2005). Higher-order substrate recognition of eIF2alpha by the RNA-dependent protein kinase PKR. *Cell* 122, 887–900.
13. Taylor, D.R., Lee, S.B., Romano, P.R., Marshak, D.R., Hinnebusch, A.G., Esteban, M., and Mathews, M.B. (1996). Autophosphorylation sites participate in the activation of the double-stranded-RNA-activated protein kinase PKR. *Mol. Cell Biol.* 16, 6295–6302.

14. Romano, P.R., Garcia-Barrio, M.T., Zhang, X., Wang, Q., Taylor, D.R., Zhang, F., Herring, C., Mathews, M.B., Qin, J., and Hinnebusch, A.G. (1998). Autophosphorylation in the activation loop is required for full kinase activity in vivo of human and yeast eukaryotic initiation factor 2alpha kinases PKR and GCN2. *Mol. Cell Biol.* 18, 2282–2297.
15. Vattem, K.M., Staschke, K.A., and Wek, R.C. (2001). Mechanism of activation of the double-stranded-RNA-dependent protein kinase, PKR: role of dimerization and cellular localization in the stimulation of PKR phosphorylation of eukaryotic initiation factor-2 (eIF2). *Eur. J. Biochem.* 268, 3674–3684.
16. Wu, S. and Kaufman, R.J. (1997). A model for the double-stranded RNA (dsRNA)-dependent dimerization and activation of the dsRNA-activated protein kinase PKR. *J. Biol. Chem.* 272, 1291–1296.
17. Dey, M., Cao, C., Dar, A.C., Tamura, T., Ozato, K., Sicheri, F., and Dever, T.E. (2005). Mechanistic link between PKR dimerization, autophosphorylation, and eIF2alpha substrate recognition. *Cell* 122, 901–913.
18. Lemaire, P.A., Lary, J., and Cole, J.L. (2005). Mechanism of PKR activation: dimerization and kinase activation in the absence of double-stranded RNA. *J. Mol. Biol.* 345, 81–90.
19. Gelev, V., Aktas, H., Marintchev, A., Ito, T., Frueh, D., Hemond, M., Rovnyak, D., Debus, M., Hyberts, S., Usheva, A., Halperin, J., and Wagner, G. (2006). Mapping of the Auto-inhibitory Interactions of Protein Kinase R by Nuclear Magnetic Resonance. *J. Mol. Biol.*
20. Wang, Z.X. and Wu, J.W. (2002). Autophosphorylation kinetics of protein kinases. *Biochem. J.* 368, 947–952.
21. George, C.X., Thomis, D.C., McCormack, S.J., Svahn, C.M., and Samuel, C.E. (1996). Characterization of the heparin-mediated activation of PKR, the interferon-inducible RNA-dependent protein kinase. *Virol.* 221, 180–188.

STRUCTURAL DYNAMIC APPROACH AS RATIONAL INPUT FOR DRUG DESIGN

ROTEM SERTCHOOK, ARIEL SOLOMON, TZVIA SELZER, AND IRIT SAGI*
Department of Structural Biology
Weizmann Institute of Science
76100 Rehovot, Israel
Corresponding author: irit.sagi@weizmann.ac.il

Abstract: Conformational flexibility and molecular dynamics play an important role in protein function. Quantification and structural characterization of protein flexibility is essential for the assignment of both the physiological and molecular mechanisms of biological macromolecules. Structural and spectroscopic tools have emerged that allow the study of local and global changes in protein structure and dynamics involved in ligand binding, protein activation, and protein–protein interactions. We have recently introduced the use of time-resolved freeze-quench X-ray absorption spectroscopy (XAS) to study detailed reaction mechanisms of metalloenzymes. Combining such measurements with molecular dynamics simulations provides novel insights into the reaction mechanism. Here we discuss the utilization of this structural dynamic approach to the design of novel inhibitors.

1. Introduction

Exploration of the mechanism of an enzymatic reaction is a long-standing objective of both experimental and computational biochemistry. Techniques that enable better understanding of the enzyme reactions mechanism and analyzing determinants of enzyme activity and specificity have the potential to contribute significantly to drug design and development. Perhaps the most obvious benefits lie in structure-based design, where mechanistic knowledge (e.g. identifying catalytic interactions at an active site) can assist in the rational design of inhibitors to the enzyme.

Examining a static picture of an enzyme, as in conventional X-ray structure analysis provides only a limited understanding of the enzyme mechanism at the atomic level of detail. In general, the catalytic cycle involves a series of intermediates and transition states, and for many of these states no detailed structural information is available. Furthermore, the position of hydrogen atoms and water molecules is often a significant feature of the enzyme mechanism, and these are in many cases unavailable from the crystallographic data. Finally, proteins and enzymes have complex dynamics, exhibiting a wide range of internal motions, some of which are vital to their function and cannot be detected by X-ray structure analysis.

The experimental characterization of catalytic states along the reaction coordinate is constantly being expanded by the advent of new experimental methods in order to provide the missing enzyme's dynamical information. Time-resolved crystallography has made remarkable improvements in the recent years [1]. However, in this method the protein's motion (conformational changes) may be restricted by the crystal packing. NMR spectroscopy

J. D. Puglisi (ed.), Structure and Biophysics – New Technologies for Current Challenges in Biology and Beyond, 111–120.
© 2007 *Springer.*

provides structural details in solution, and can detect equilibrium fluctuations on timescales of millisecond and longer [2]. Study of metalloenzymes dynamic features is possible in a similar timescale by the use of time-resolved freeze-quench X-ray absorption spectroscopy (XAS) coupled with transient kinetic assay as introduced recently by Sagi and coworkers [3]. Earlier times are accessible to optical spectroscopic techniques such as electronic absorption, emission, and circular dichroism (CD) spectroscopy, as well as infrared and Raman spectroscopy [4].

A better understanding of the dynamic nature of enzymatic reactions opens up a novel path to rational drug design by targeting inhibitors toward dynamic protein conformations and not just by considering "static" structural features.

2.   Structure-Function Analysis of Highly Homologous Enzymes

The challenge of elucidating an enzyme's mechanisms based solely on its structure is further enhanced in families of homologous enzymes. In such cases, the similarity in the sequence is expressed in a similar structure that can be further interpreted as having a similar mechanism of catalysis and similar susceptibility to various inhibitors. Yet, even in such families of homologous enzymes, each enzyme might be unique in terms of its substrate specificity and inhibition profile, indicating that additional factors other than the active site structure must be significant for controlling the enzyme's catalytic mechanism.

The search for additional factors that govern individual enzyme activity is of special interest in the designing of new drug candidate that should inhibit exclusively single enzyme. In general, inhibitors used in therapy must show specificity toward the target enzyme. Inhibition of related enzymes with a different biological role will lead usually to severe side effects. In fact, it is estimated that 10% of the drug candidates ultimately failed in clinical development because of side effects [5], which many times reflect insufficient specificity of the inhibitor to its target enzyme.

Comparative structural-spectroscopic analysis conducted on the TNF-converting enzyme (TACE or ADAM-17) and the matrix metalloproteinases (MMPs) demonstrates that highly homologous catalytic sites exhibit diverse biophysical characteristics. These cancer-associated proteases exhibit remarkable structural similarity but different biological role. TACE is endopeptidase that cleaves a specific site in the membrane-bound precursor of tumor necrosis factor α (TNF-α) to its mature soluble form [6, 7]. It is a zinc-dependent metalloproteinase belonging to the astacin/adamalysin family (M12, according to MEROPS classification [8]). Human MMPs constitute a multigene family of about 30 secreted soluble and membrane-tethered enzymes that cleave a variety of pericellular substrates [9]. The MMPs are structurally similar zinc-dependent endopeptidases belonging to the M10 metalloproteinase family.

The sequence alignment of TACE with members of the MMP family indicates substantial similarity, particularly within the catalytic domain (35–44% identity). The high degree of sequence conservation between TACE and the MMPs translates into substantial structural similarity. Figure 1A shows the overlapping structures of the catalytic domains of TACE, MMP-2, and MMP-9. The core structures of these enzymes are highly similar, varying mostly within the peripheral loops. Moreover, the three-dimensional structural elements surrounding the zinc-binding site in these enzymes are almost identical including the zinc ligands and catalytic glutamate in the consensus motif **HEXXHXXGXXH** (Figure 1B). TACE and the MMPs also share remarkable structural similarity of the substrate-binding hydrophobic pocket [10]. This subsite similarity to the MMPs explains the observed sensitivity of a TACE-like activity to synthetic hydroxamic acid inhibitors originally designed for inhibition of various MMPs [11]. Consequently, because of the high structural similarities between the active site of TACE and related zinc endopeptidases such as MMPs the design of potent as well as selective inhibitors for TACE is not trivial.

To obtain new insights into factors that govern each individual enzyme, Solomon et al. [12] used a selective MMP inhibitor as a probe to examine the structural and kinetic effects occurring at the active site of TACE upon inhibition. They used the selective MMP mechanism-based inhibitor SB-3CT [13] to characterize the fine structural and electronic differences between the catalytic zinc ions within the active sites of TACE and MMP-2. The experimental arsenal included XAS far-UV CD and steady-state kinetic analysis. It was found that SB-3CT directly binds the metal ion of TACE as observed in MMP-2. However, in contrast to MMP-2, the binding mode of SB-3CT to the catalytic zinc ion of TACE differs in bond distance and in the total effective charge of the catalytic zinc ion. Furthermore, SB-3CT was found to inhibit TACE in a noncompetitive fashion by inducing significant conformational changes in the structure, as can be seen in the CD spectrum (Figure 2). For MMP-2, SB-3CT behaved as a competitive inhibitor and no significant conformational changes were observed (Figure 2).

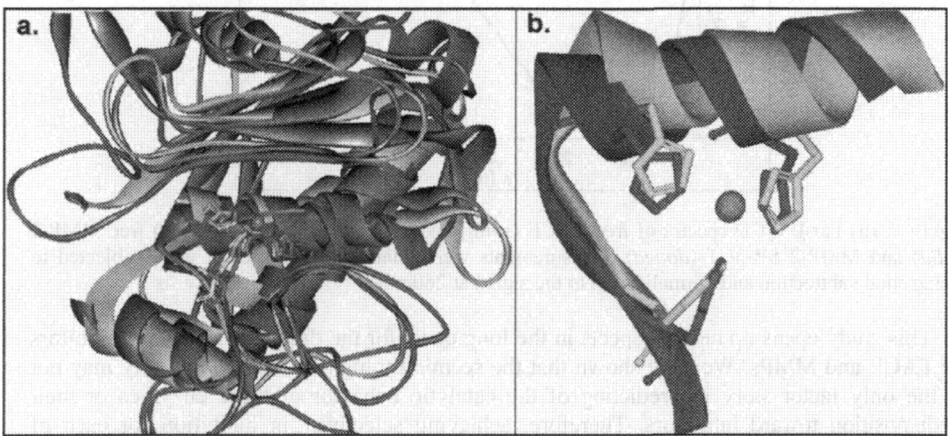

Figure 1. (**a**) Structural overlap of the catalytic domains of TACE (PDB code 1BKC, pink), MMP-2 (PDB code 1CK7, *yellow*), and MMP-9 (PDB code 1L6J, *gray*).The catalytic zinc ion is depicted as an *orange ball*. (**b**) A closer view of the catalytic zinc ion of TACE (*pink*) and MMP-2 (*gold*) shows remarkable structural conservation at the proximal environment of the metal ion (*orange ball*).

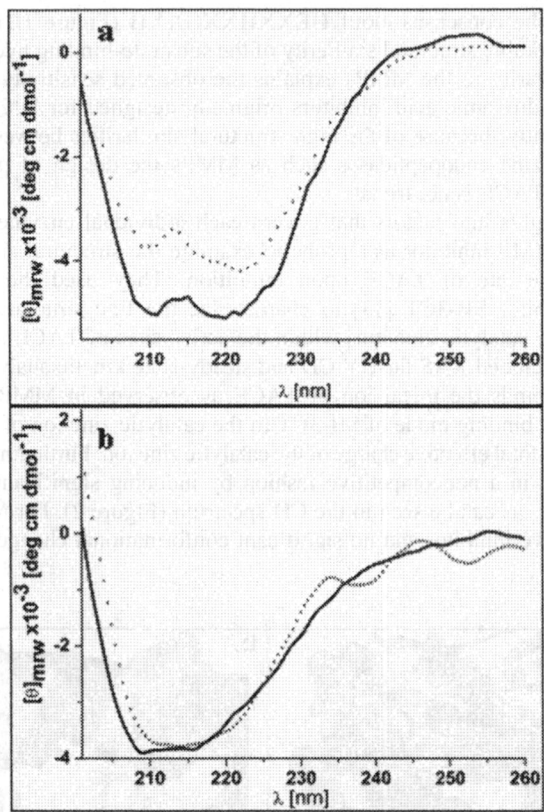

Figure 2. (**a**) Far-UV CD spectra of free TACE (*solid*) and TACE SB-3CT (*dotted*) (**b**) free MMP-2 (*solid*) and MMP-2 SB-3CT (*dotted*). Measurements were done at 25°C. Spectra were subjected to background subtraction and normalization to the signal at 260 nm.

This study opens up new prospects in the long quest for the design of selective inhibitors for TACE and MMPs. We have shown that the seemingly high structural similarity may not be the only factor story in predicting of the catalytic behavior of these enzymes or their predisposition toward inhibitors. Therefore, achieving selectivity in inhibition for each of these enzymes should prove feasible if a multidisciplinary approach would venture to consider factors other than only the static structural features. Hence, the development of new structural dynamic experimental approaches is of critical importance.

3.   Dynamic Structure Function Investigation

Probing metalloenzyme reactions by in situ XAS is a promising experimental method for explaining the changes that occur in critical metal centers during the course of the enzymatic reaction [14–19]. Specifically for catalytic zinc sites, XAS is the only spectroscopic method that can provide high-resolution structural and electronic information in solution. Sagi and coworkers developed the time-resolved freeze-quench XAS technique to determine the intermediate states at the first coordination shell of the catalytic metal ion during substrate turnover. This dynamic structure-function investigation of metalloenzymes tightly links structure to function by determining the reaction time course using steady-state kinetics followed by investigating the kinetic mechanism with a millisecond timescale by pre-steady-state kinetics using a stopped flow machine.

The online monitoring of the kinetics can be done using any spectroscopic measurement such as UV/VIS, CD, fluorescence emission, and others. The kinetic trace can indicate the different kinetic phases that occur during the reaction and point out possible intermediates that evolve during a single catalytic cycle. A freeze-quench module of the stopped-flow machine is used as follows: reaction components are rapidly mixed (dead time ~2 ms) and sprayed into a pre-cooled isopentane bath and collected inside a sample holder. Enzyme-substrate complexes formed at different reaction times are trapped and collected according to the kinetics and analyzed by XAS (see experimental setup in Figure 3).

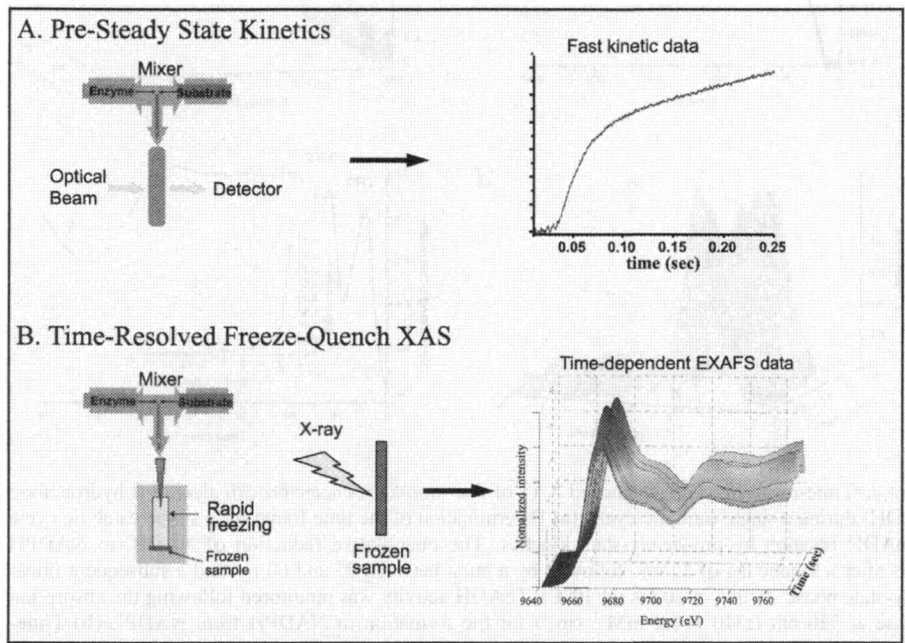

Figure 3. Structural dynamic investigation setup: A biological reaction catalyzed by a metalloenzyme is measured using pre-steady-state kinetics of any spectroscopic method (**A**). The kinetic trace is analyzed and reaction components are mixed and freezed accordingly (**B**). The time-dependent frozen samples are measured using XAS and data processing and analysis provides time-resolved structural information about the reaction.

XAS provides local structural and electronic information about the nearest coordination environment surrounding the catalytic metal ion within the active site of a metalloprotein in solution. The X-ray absorption coefficient data of various time-dependent complexes can be collected at any K-edge energy according to the metal type. Trapping enzyme-substrate complexes during single catalytic turnover increases the probability of trapping reaction intermediates [3]. Distinct changes in spectral features can be observed in the raw XAS spectra. These changes are even more apparent in the radial distribution of the raw extended X-ray absorption fine structure (EXAFS) data as represented by the Fourier transform (FT) spectra of the various time domains (see Figure 4 from Kleifeld et al.) [3]. The shape

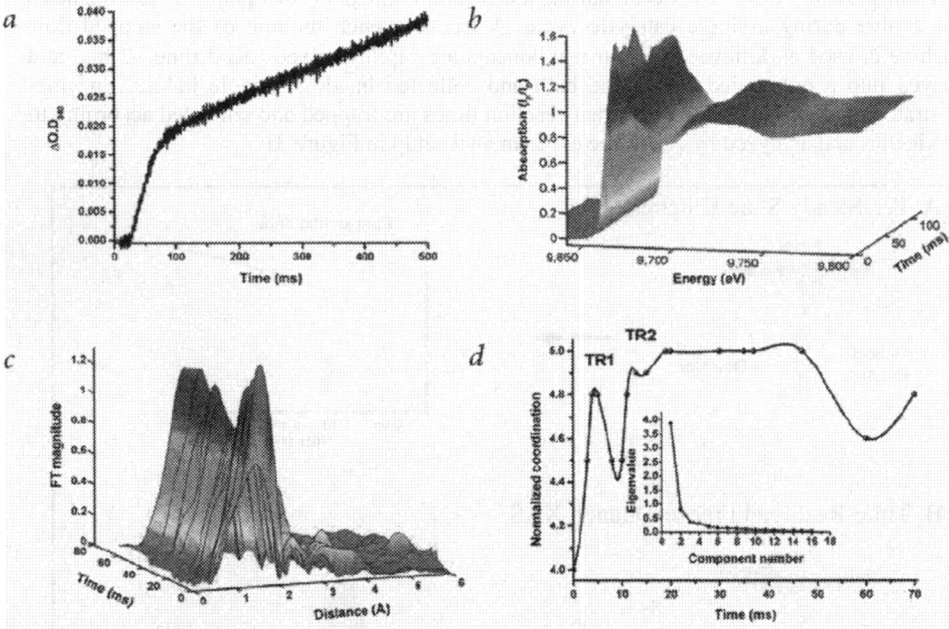

Figure 4. Time-resolved freeze-quenched XAS of *Thermoanaerobacter brockii* alcohol dehydrogenase (TbADH) during a single catalytic cycle. (**a**) Determination of the time frame for a single catalytic cycle of TbADH reaction by pre-steady-state kinetics. The quantitative reduction of NADP$^+$ to NADPH occurs after a kinetic lag of 25 ms, followed by a burst between 25 and 60 ms and a subsequent linear steady-state phase starting at 70 ms (at 10°C). TbADH activity was monitored following the absorption increase at 340 nm ($\varepsilon 340 = 6.2$ mM^{-1} cm^{-1}) for the formation of NADPH from NADP$^+$. (**b**) Time-resolved freeze-quenched raw X-ray fluorescence data. Spectral changes can be observed within the time frame of single catalytic cycle (0–60 ms). (c) Magnitudes of the Fourier transform (FT) of k^2-weighted spectra, uncorrected for the photoelectron phase shifts, of the various time domains representing changes in the radial distribution of the zinc–ligand first shell environment during catalysis. (**d**) The change in coordination number during catalysis of TbADH. The coordination of the zinc ion expands from its original tetrahedral structure to the pentacoordinate complex by the addition of Zn–N/O contribution. Two distinct pentacoordinate intermediates, TR1 and TR2, evolve during the time frame of the kinetic lag phase. The normalized total coordination number is calculated by the relative fractions obtained from the RPA EXAFS analysis for each time point. Insert, the PCA "screen test" results demonstrating that all the time-dependent EXAFS spectra can be reproduced using two principal components.

and amplitude of the first FT peak are directly related to the details of the atomic distribution of the amino acid residues within the first coordination shell. The variations in the first FT peak indicate that the local structure around the catalytic metal ion undergoes distinctive dynamic structural changes during turnover. Detailed data analysis involves the establishment of the local structure around the catalytic metal ion within the enzyme active site by (i) obtaining the number of different intermediate species in all time phases, (ii) identifying those species and (iii) defining their relative abundance. This is achieved by a novel combination of principal component analysis (PCA), multiple data-set fits (MDS) and residual phase analysis (RPA) [3].

4.   Application of the Dynamic Structure-Function Investigation to the Design of Inhibitors

Dynamic structure-function investigations such as stopped-flow XAS conjugated with transient kinetics can reveal chemical, electronic, and conformational diversity along the reaction pathway even in closely related enzymes. These differences between the enzymes can be further exploited for rational drug design of potent and selective inhibitors for the individual enzymes.

An example of such differences between structurally similar enzymes might be the polarity of the amino acid residues in the second shell around the metal center. This subtle change can affect the metal's effective charge, can further induce changes in the pKa of incoming inhibitors and eventually affect the inhibitor binding mode [20]. In the case of the structurally similar enzymes TACE and the MMP-2, the small variation in the polarity of the active site residues of the catalytic zinc was interpreted as the main factor that caused the differences in the binding of the mechanism-based inhibitor SB-3CT to both enzymes. Combining the XAS results and theoretical modeling reveals possible binding modes of the inhibitor in TACE and MMP-2, as can be seen in Figure 4 [12].

The finding of different possible binding modes in the extended tunnel of TACE active sites is an example of secondary subsites within the active site. Such subsites can be exploited for the design of new potent and selective inhibitors of TACE by targeting the designed inhibitor to the specific subsites at the vicinity of the catalytic zinc ion. In general, secondary subsites have the advantage of possible specificity to individual enzymes together with possible potency owing to the interaction with the significant catalytic residues.

Another difference between structurally similar enzymes could be the transition state interaction in the active site. Basically, the enzymatic reaction rate acceleration as well as its selectivity is rationalized by selective stabilization of the transition state by strong and preferential binding of the enzyme to the transition state. Therefore a key to understanding enzyme catalysis is to define how enzymes recognize substrates in the transition state, where the active-site amino acid residues are assembled to interact specifically with activated forms of the substrates to promote the chemical reaction. Consequently, mimicking the transition-state's structural and electronic features is a general and rational method for the molecular design of potent and selective inhibitors known as transition state analogs (TSAs). Transition states are by definition extremely unstable species (lifetimes of $\sim 10^{-13}$ s), with only partially formed bonds. As such, TSAs usually mimic high-energy intermediates, and not real transition states. Nevertheless, even the partial mimic of transition states is capable of making potent inhibitors as was shown for many TSAs [21, 22].

a.

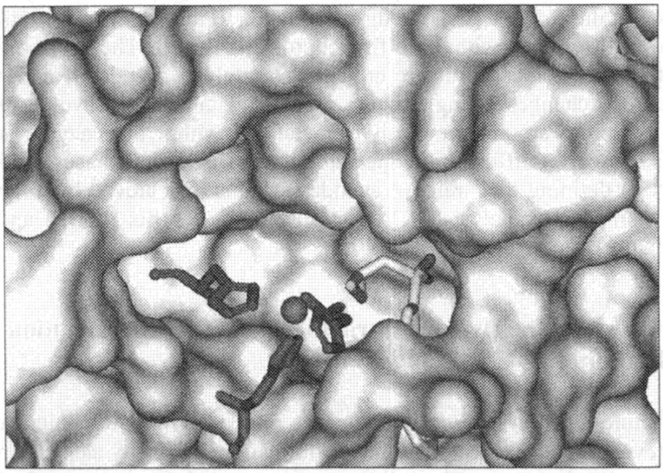

b.

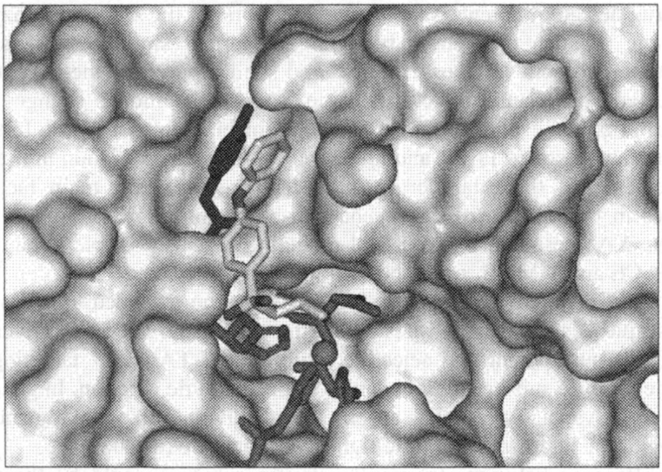

Figure 5. Dynamic modeling of the catalytic zinc site in the (**a**) MMP-2·SB-3CT complex and (**b**) TACE·SB-3CT complex. The active site is represented in stereo as a van der Waals surface in *gray* and the catalytic zinc ion is shown as an *orange* sphere. The inhibitor molecule and the zinc ligands are shown in a ball-and-stick representation (carbon in *white*; oxygen in *red*; and sulfur in *yellow*). In the case of MMP-2·SB-3CT complex the biphenyl moiety of this mechanism-based inhibitor fits the P1 (arrowhead) site and the thiirane ring is at the expected position to coordinate the zinc ion.

By the use of multidisciplinary dynamic structure-function investigations, we aim to capture high-energy conformational transitions along the reaction coordinate rather than the low-energy ground state conformers, which possess a high degree of similarity. Consequently, having access to transition states and using this data as input to rational drug design, provides a very powerful way of designing highly potent and selective drugs.

5. Conclusions

Enzymes are inherently flexible therefore, it is possible that protein motions couple to the formations of transition state like conformations, influence the probability of a reaction to occur [23]. Protein structures associated with these conformational transitions along a reaction pathway must be quantified if the structural basis of function is to be fully understood. The exploration for links between protein structure, conformation, and catalysis is constantly being expanded by the advent of new theoretical and experimental methods, including, optical spectroscopic techniques such as electronic absorption, emission, and CD spectroscopy. Recently we have introduced the use of time-resolved freeze-quench XAS coupled with a transient kinetic assay to the study of metalloenzymes reactions during enzyme turnover. These experiments provide detailed structural and chemical characterization of catalytic metal sites in real-time. Here we argue that better understanding of the dynamic nature of enzymatic reactions opens up a novel path to rational drug design by targeting inhibitors toward dynamic protein conformations and not just by considering "static" structural features.

References

1. Schotte, F., Lim, M.H., Jackson, T.A., Smirnov, A.V., Soman, J., Olson, J.S., Phillips, G.N., Jr., Wulff, M., and Anfinrud, P.A. (2003). Watching a protein as it functions with 150-ps time-resolved X-ray crystallography. *Science* 300, 1944–1947.
2. Dyson, H.J. and Wright, P.E. (2004). Unfolded proteins and protein folding studied by NMR. *Chem. Rev.* 104, 3607–3622.
3. Kleifeld, O., Frenkel, A., Martin, J.M., and Sagi, I. (2003). Active site electronic structure and dynamics during metalloenzyme catalysis. *Nat. Struct. Biol.* 10, 98–103.
4. Balakrishnan, G., Case, M.A., Pevsner, A., Zhao, X., Tengroth, C., McLendon, G.L., and Spiro, T.G. (2004). Time-resolved absorption and UV resonance Raman spectra reveal stepwise formation of T quaternary contacts in the allosteric pathway of hemoglobin. *J. Mol. Biol.* 340, 843–856.
5. Dimasi, J.A. (2001). Risks in New Drug Development: Approval Success Rates for Investigational Drugs. *Clin. Pharmacol. Ther.* 69, 297–307.
6. Moss, M.L., Jin, S.L., Milla, M.E., Bickett, D.M., Burkhart, W., Carter, H.L., Chen, W.J., Clay, W.C., Didsbury, J.R., Hassler, D., Hoffman, C.R., Kost, T.A., Lambert, M.H., Leesnitzer, M.A., McCauley, P., McGeehan, G., Mitchell, J., Moyer, M., Pahel, G., Rocque, W., Overton, L.K., Schoenen, F., Seaton, T., Warner, J., Willard, D., Su, J.L., and Becherer, J.D. (1997). Cloning of a disintegrin metalloproteinase that processes precursor tumour-necrosis factor-alpha. *Nature*, 385, 733–736.
7. Black, R.A., Rauch, C.T., Kozlosky, C.J., Peschon, J.J., Slack, J.L., Wolfson, M.F., Castner, B.J., Stocking, K.L., Reddy, P., Srinivasan, S., Nelson, N., Boiani, N., Schooley, K.A., Gerhart, M., Davis, R., Fitzner, J.N., Johnson, R.S., Paxton, R.J., March, C.J., and Cerretti, D.P. (1997). A metalloproteinase disintegrin that releases tumour-necrosis factor-alpha from cells. *Nature*, 385, 729–733.
8. Rawlings, N.D., Tolle, D.P., and Barrett, A.J. (2004). MEROPS: the peptidase database. *Nucleic Acids Res.* 32(Database issue), D160–164.
9. Nagase, H. and Woessner, J.F., Jr. (1999). Matrix metalloproteinases. *J. Biol. Chem.* 274, 21491–21494.
10. Maskos, K., Fernandez-Catalan, C., Huber, R., Bourenkov, G.P., Bartunik, H., Ellestad, G.A., Reddy, P., Wolfson, M.F., Rauch, C.T., Castner, B.J., Davis, R., Clarke, H.R., Petersen, M.,

Fitzner, J.N., Cerretti, D.P., March, C.J., Paxton, R.J., Black, R.A., and Bode, W. (1998). Crystal structure of the catalytic domain of human tumor necrosis factor-alpha-converting enzyme. *Proc. Natl. Acad. Sci. USA*. 95, 3408–3412.

11. DiMartino, M., Wolff, C., High, W., Stroup, G., Hoffman, S., Laydon, J., Lee, J.C., Bertolini, D., Galloway, W.A., Crimmin, M.J., Davis, M., and Davies, S. (1997). Anti-arthritic activity of hydroxamic acid-based pseudopeptide inhibitors of matrix metalloproteinases and TNF alpha processing. *Inflamm. Res*. 46, 211–215.

12. Solomon, A., Rosenblum, G., Gonzales, P.E., Leonard, J.D., Mobashery, S., Milla, M.E., and Sagi, I. (2004). Pronounced diversity in electronic and chemical properties between the catalytic zinc sites of tumor necrosis factor-alpha-converting enzyme and matrix metalloproteinases despite their high structural similarity. *J. Biol. Chem*. 279, 31646–31654.

13. Brown, S., Bernardo, M.M., Li, Z.H., Kotra, L.P., Tanaka, Y., Fridman, R., and Mobashery, S. (2000). Potent and selective mechanism-based inhibition of Gelatinases. *J. Am. Chem. Soc*. 122, 6799–6800.

14. Powers, L., Sessler, L., Woolery, G.L., and Chance, B. (1984). CO bond angle changes in photolysis of carboxymyoglobin. *Biochemistry* 23, 5519–5523.

15. Scheuring, E.M. Clavin, W., Wirt, M.D., Miller, L.M., Fischetti, R.F., Lu, Y., Mahoney, N., Xie, A., Wu, J., and Chance, M.R. (1996). Time-resolved X-ray absorption spectroscopy of photoreduced base-off Cob(II)alamin compared to the Co(II) species in Clostridium thermoaceticum. *J. Phys. Chem*. 100, 3344–3348.

16. Chance, M.R., Miller, L.M., Fischetti, R.F., Scheuring, E., Huang, W.X., Sclavi, B., Hai, Y., and Sullivan, M. (1996). Global mapping of structural solutions provided by the extended X-ray absorption fine structure ab initio code FEFF 6.01: structure of the cryogenic photoproduct of the myoglobin-carbon monoxide complex. *Biochemistry* 35, 9014–9023.

17. Riggs-Gelasco, P.J., Shu, L., Chen, S., Burdi, D., Huynh, B.H., Que, L., Jr., and Stubbe, J. (1998). EXAFS characterization of the intermediate X generated during the assembly of the E. coli ribonucleotide reductase R2 diferric tyrosyl radical cofactor. *J. Am. Chem. Soc*. 120, 849–860.

18. Hwang, J., Krebs, C., Huynh, B.H., Edmondson, D.E., Theil, E.C., and Penner-Hahn, J.E. (2000). A short Fe-Fe distance in peroxodiferric ferritin: control of Fe substrate versus cofactor decay? *Science* 287, 122–125.

19. Lee, S.K., George, S.D., Antholine, W.E., Hedman, B., Hodgson, K.O., and Solomon E.I. (2002). Nature of the intermediate formed in the reduction of O2 to H2O at the trinuclear copper cluster active site in native laccase. *J. Am. Chem. Soc*. 124, 6180–6193.

20. Cross, J.B., Duca, J.S., Kaminski, J.J., and Madison, V.S. (2002). The active site of a zinc-dependent metalloproteinase influences the computed pKa of ligands coordinated to the catalytic zinc ion. *J. Am. Chem. Soc*. 124, 11004–11007

21. Schramm, V.L. (1998). Enzymatic transition states and transition state analog design. *Annu. Rev. Biochem*. 67, 693–720.

22. Robertson, J.G. (2005). Mechanistic basis of enzyme-targeted drugs. *Biochemistry*. 44, 5561–5571.

23. Benkovic, S.J. and Hammes-Schiffer, S. (2003). A perspective on enzyme catalysis. *Science*. 301, 1196–1202.

MAX PERUTZ: CHEMIST, MOLECULAR BIOLOGIST, AND HUMAN RIGHTS ACTIVIST[†]

JOHN MEURIG THOMAS
Department of Materials Science
University of Cambridge,
Cambridge CB2 3QZ, UK and
Davy Faraday Research Laboratory
Royal Institution, London W1S 4BS

In tracing the trajectory of Max Perutz's life, future historians of science will doubtless highlight several great scientific adventures and achievements:

(i) He founded, with Sir Lawrence Bragg and John Kendrew, the Medical Research Council (MRC) Unit of Molecular Biology in the Cavendish Laboratory, Cambridge, in 1947, and later he became the principal scientific architect of the Laboratory of Molecular Biology (LMB) which he founded in Cambridge in 1962.

(ii) Along with his associate, John Kendrew, he solved the first protein structures[1] (haemoglobin and myoglobin), and this earned them the Nobel Prize in Chemistry in 1962.

(iii) Again, with John Kendrew, he founded the European Molecular Biology Organisation (EMBO), and became its founding chairman in 1963.

(iv) By focusing on numerous mutants of haemoglobin, from a large range of living creatures and numerous humans, he gained a deep understanding of several inherited diseases enabling him to open up the new field of molecular pathology and adding to our knowledge of molecular evolution. He elucidated the nature of such tragic diseases as thalassemia and sickle-cell anaemia.

(v) In 1970, he finally worked out the mode of action of haemoglobin[2] and, in 1986, nearly a quarter of a century after his Nobel Prize-winning work, he discovered how haemoglobin acts as a drug receptor.

(vi) As Francis Crick wrote in 2002,[3] Max Perutz was still the centre of the revolution in molecular biology that occupied the second half of the twentieth century.

And the careful historian of science will also record that, in 1948, the 34-year-old Perutz solved the problem of how a glacier flows. (It moves, not like treacle, but more like a ductile metal when it is extended, with planes of atoms gliding over one another.)

All these, and many other scientific achievements, are associated with Max Perutz's name. But to those who knew him, to those who worked or lived alongside him, to those who observed his quiet, effective negotiating skills, and to those who had the pleasure of talking to or corresponding with him, or attending his lectures, or of reading his evocative book reviews, essays, and letters, there was far more to Max Perutz. He combined, in a singular fashion, all the noblest instincts of mankind.

Max Perutz was a man of warm humanity and of great human decency and compassion. He had immense moral courage. He was morally incorruptible. And he possessed huge reserves of intellectual energy, as well as a youthful voracity for new knowledge. He was a

J. D. Puglisi (ed.), Structure and Biophysics – New Technologies for Current Challenges in Biology and Beyond, 121–126.
© 2007 *Springer.*

stylish and incisive author of popular scientific articles and reviewer of books – books that he meticulously researched and fastidiously, though eloquently, analysed. He wrote charming and sensitive personal letters. Above all, he was an indefatigable warrior, passionately committed to social and political justice. Intellectual honesty and freedom, and especially human rights mattered to him profoundly.

Max Perutz often exhibited the temperament of the artist and the imaginative sensibility of the poet. It pleased his many admirers, and Max himself, when Rockefeller University accorded him their first Lewis Thomas Prize, recognising the Scientist as Poet.

Max delighted in the beauty of the natural world. He was the kind of man who, before starting his laboratory work at the LMB on a Spring morning, would occasionally take a walk on the Gog Magog hills (outside Cambridge) filling his heart and soul, in so doing, with pantheistic pleasure.

But Max was resolute in his opposition to what he perceived to be wrong-headed and erroneous arguments or decisions. Long before his work at Cambridge came to fruition – long before he made his monumental scientific breakthroughs – he felt impelled to resign from his post as lecturer in the University of Cambridge, as a protest against the decisions of the central authorities.

Another example of how forthright he could be is seen in his attack on certain philosophers and historians of science whose theses he disputed. Max rejected as nonsense the view, popular among modern sociologically oriented philosophers of science, that scientific truth is relative and shaped by a scientist's personal concerns, including his or her political, philosophical, even religious instincts. When he attacked such opinions, he once quoted Max Planck's memorable assertion:

> *There is a real world independent of our senses: the laws*
> *of nature were not invented by man, but forced upon him*
> *by that natural world. They are the expression of a rational*
> *order.*

Max would probably have agreed with Richard Feynman's flippant remark:

> *Philosophers of science are about as helpful to scientists*
> *as ornithologists are to birds.*

Max's long, labyrinthine path as a research scientist began when he studied chemistry at the University of Vienna, his home city. He acquired a special interest in organic bio-chemistry and heard about the work of Sir Gowland Hopkins, the discoverer of vitamins. Max decided that he wanted to solve a great problem in biochemistry. His teacher, Hermann Mark, visited Cambridge and had planned to pave the way for Max to join Hopkins' group there. But Mark met J.D. Bernal, a pyrotechnically brilliant conversationalist, who said he would take Max as his student. (Mark forgot to approach Hopkins!) So, in 1936, Max became a researcher in the Cavendish Laboratory where Bernal taught and researched in physics, and a graduate student at Peterhouse.

On Bernal's advice, he learned X-ray crystal structure analysis in the Department of Mineralogy. A year or so later, he visited his cousin Felix Horauwiz (in Prague) who convinced him that an appropriate target for his ambitions was the structure of haemoglobin, first, because it was the protein that was most abundant and easiest to crystallise, second, because oxyhaemoglobin and deoxyhaemoglobin had different crystal structures – but no one knew what these structures were. Gradually, it emerged that each unit repeat volume in a crystal of haemoglobin has about 12,000 atoms. In 1937, when Max made his decision, X-analysis had solved structures containing no more that about 100 atoms. *That* was the magnitude of the problem Max set himself.

He had been encouraged, however, by the success J.D. Bernal and Dorothy Crowfoot (later Hodgkin) had achieved in obtaining in 1934 beautiful X-ray diffraction patterns of the protein crystal, pepsin, in its mother liquor. Soon, he, Bernal and Fankuchen obtained[4] similarly encouraging diffraction patterns from haemoglobin and chymotrypsin. But it was not until the late 1950s, under the aegis of Sir Lawrence Bragg, that he finally reached his target of elucidating the structure of haemoglobin. And when he did, it made him famous. From 1936 to the late 1950s, however, he suffered a succession of setbacks: there were many scientific, personal and political obstacles to surmount. In 1940, his studies at the Cavendish Laboratory were rudely interrupted by his internment (along with hundreds of German-speaking people then living in the UK) first in the Isle of Man, then in Quebec, Canada. He returned to work of national importance during the war. In 1942, after a whirlwind romance, he married Gisela Peiser, a Berlin-born lady then working in Cambridge; and in 1943 he became a British citizen. In 1944, he was back again at the bench in the "Cavendish", where, in 1945, he was joined by John Kendrew. Francis Crick, a physicist, joined the group as Max's Ph.D. student in 1948. Jim Watson, a geneticist, came in 1951 and was soon working with Crick on DNA.

In early 1951, after about 6 years of extracting what X-ray crystallographers call Patterson maps (which, in the case of haemoglobin crystal, consisted of some 25 million lines between the thousands of atoms in the haemoglobin molecule), Max Perutz felt elated when they seemed to tell him that haemoglobin consists simply of bundles of parallel chains of atoms spaced apart at equal intervals. I quote his words:

> *Shortly after my results appeared in print, a new graduate student joined me. As his first job, he performed a calculation which proved that no more than a small fraction of the haemoglobin molecule was made up of the bundles of parallel chains that I had persuaded myself to see, and that my results, the fruits of years of tedious labour, provided no other clue to its structure. It was a heartbreaking instance of patience wasted, an ever-present risk in scientific research.*

That graduate student made himself unpopular in the MRC unit of the Cavendish at the time. But he was very clever. In fact, years later, Max Perutz told me that that student turned out to be one of the cleverest men he ever met. His name was Francis Crick – a man who won the Nobel Prize, with Watson and Wilkins, before he completed his PhD!

After a period of deep depression, which disturbed Max emotionally and physically, a ray of brilliant light appeared in 1953. Max, remembering an earlier suggestion by Bernal, realised that he could benefit by tagging molecules of haemoglobin with heavy ions such as silver or mercury. Being the expert crystallographer that he was, he knew immediately that such heavy-atom-tagging should enable him to solve the structure of haemoglobin in a manner quite different from his early approach, which Francis Crick had so comprehensively and unceremoniously demolished. Both Perutz and Kendrew redoubled their efforts. Max it was who first demonstrated the validity of the method, by computing the X-ray diffraction patterns of haemoglobin with and without a mercury tag. (Sir Lawrence Bragg was so thrilled that, to quote Max, he "went around telling everyone that I had discovered a goldmine".)

But John Kendrew, in 1958, working both at the Cavendish and with David Phillips at the Royal Institution, solved the three-dimensional structure of myoglobin, an achievement greeted worldwide as sensational. Max was both pleased and somewhat depressed with this breakthrough. Pleased because his method and his laboratory and his partner, John Kendrew, had triumphed. But he said later that he was also depressed, partly because he had not "got" to haemoglobin first, but partly also because he had a nagging uncertainty that the solution of the haemoglobin problem might prove bewilderingly and interminably elusive. In September 1959, however, Max Perutz and his colleagues, using 40,000 measurements from crystals of haemoglobin and six heavy-atom derivatives, calculated the three-dimensional structure of the molecule. At last, he had reached the longed-for shore.

Max officially retired from the LMB in 1979, but he worked there almost every day up until the time of his death in 2002. And only a few days before he entered hospital during his terminal illness, he completed the text of a research article that followed on from his important work on the fundamental causes and molecular aspects of neurodegenerative diseases.

It is universally acknowledged that the LMB is one of the most famous and successful research laboratories now in existence. Max had set up a simple structure for running the LMB from its inception in 1962. "I persuaded the MRC" he said "to appoint me as Chairman of the Governing Board rather than Director, a Board to be made up of Kendrew, Crick, Sanger and me" (four wise men, five Nobel Prizes!). "This arrangement reserved major decisions of scientific policy to the Board and left their execution to me. The Board met only rarely!

Shortly after he passed away in 2002, I discussed elsewhere[5] the scientific and humane legacy of Max Perutz. In particular, I sought to divine the secret of the extraordinary success of the LMB, and to contrast his methods of running a research laboratory with the advice nowadays given to scientists by the Paladins of accountability in various Funding and Research Councils, and increasingly by university administrators. The principles he used were:

> choose outstanding people and given them intellectual freedom; show genuine interest in everyone's work, and give younger colleagues public credit; enlist skilled support staff who can design and build sophisticated and advanced new apparatus and instruments; facilitate the interchange of ideas, in the canteen as much as in seminars; have no secrecy; be in the laboratory most of the time and accessible to everybody where possible; and engender a happy environment where people's morale is kept high.

These are lofty principles, obviously and compellingly correct, but difficult to live up to. A crystallographer friend of mine, who visited me recently, said of them that they reminded him of the "Sermon on the Mount" or the "Declaration of Independence". Max, however, complied with these principles, and he was ably assisted for many years by his devoted wife, Gisela, who made the canteen of the LMB a focal point of intellectual stimulus.

My friendship with Max extended over the last 24 years of his life: we lived a few doors from one another; we were members of the same Cambridge college, Peterhouse; and for part of that time I had responsibilities for running the Royal Institution and the Davy Faraday Research Laboratory, places where he and John Kendrew had been Readers for 13 years from the time of the appointment of my predecessor-but-one, Sir Lawrence Bragg, as Director.

Through my friendship with Max, I benefited enormously from his wisdom, guidance, and humour, which I grew to appreciate during our numerous walks around the playing fields adjacent to our homes, while strolling in the Botanical Garden, or sitting for tea in the intimacy of our homes. During those discussions, I recall particularly two anecdotes worthy of narrating here. The first relates to an incident that occurred while he attended a "Human Rights" gathering. A Soviet scientist had said that one should cease to use the term "freedom of speech" and replace if with "freedom after speech". The second involves his retort when I asked him how he had become such a skilled negotiator. He replied by quoting what a former Fellow of Trinity College, Cambridge had once said:

> In Cambridge, to reach your goal, you must learn to combine the linear persistence of the tortoise with the circuitous locomotion of the hare.

Max was utterly repulsed by the thought of the use of torture on political or other prisoners. He could be seen to cringe while talking about it. His revulsion of such practices was partly what animated him as a human rights activist. But he detested injustice of any kind, and was dedicated to the eradication of ignorance. He did something about it. Members of this audience will know that 10 years ago, in Amsterdam at the Dutch Academy, he read a paper on "By What Right Do We Invoke Human Rights?" This widely published lecture[6] is a closely reasoned history of the concept of human rights from the days of Aeschylus (458 BC) to the present day. His response to the terrorist attack in New York on 9/11 was to organise a petition intended for world leaders. Amongst other things it said, "Avoid military actions against innocent people. Military retaliation does not solve the problem of fanaticism, but instead fuels the anger by demanding 'counter' revenge."

In closing this tribute, having heard repeated mention today of liberty, freedom, the pursuit of truth, and the elimination of injustice, I can think of no better way to remember Max, and to remind us of the things that he stood for, than to quote some of the words of the Hindu mystic and poet, Rabindranath Tagore (Gisela, Max's wife, had met Tagore in Berlin). Tagore and Einstein had an interesting correspondence about 90 years ago. Tagore held that scientific truth was realised through man, whereas Einstein maintained (as did Max Planck, whom I quoted earlier) that scientific truth must be conceived as a valid truth that is independent of humanity.

Knowing that the premier academics and scholarly bodies of the world are committed to the restless pursuit of truth and knowledge (as Max was), it is appropriate that I should recite, to end, Song 35 of Tagore's "Gitanjali":

> Where the mind is without fear and the head is held high;
> Where knowledge is free;
> Where the world has not been broken up into fragments by narrow domestic walls;
> Where words come out from the depth of truth;
> Where tireless striving stretches its arms towards perfection;
> Where the clear stream of reason has not lost its way into the dreary desert
> sand of dead habit;
> Where the mind is led forward by thee into ever-widening thought and action
> Into that heaven of freedom, my Father, let my country awake.

References

1. The protein structure data base in the USA now contains over thirty thousand structures, essentially all derived using the crystallographic method pioneered by Perutz and Kendrew.
2. The words used by him were: *"Haemoglobin is not just an oxygen tank: it is a molecular lung. It changes its structure every time it takes up and releases oxygen. You can hear your heart going 'thump, thump, thump'; but in your blood the haemoglobin molecules go 'click, click, click' – but you can't hear that."*
3. Francis, C., *Physics Today* (2002). Aug.
4. Thomas, J.M., *Angew Chemie. Intl. Ed. Eng.* (2002). 41, 3155
5. Bernal, J.D., Fankuchen, I., and Perutz, M.F. *Nature* (1938) 141, 523
6. Perutz, M.F. *Proc. Am. Philos. Soc.* (1996). 140, 135.

RIBOSOMAL CRYSTALLOGRAPHY: PEPTIDE BOND FORMATION, CHAPERONE ASSISTANCE, AND ANTIBIOTICS INACTIVATION

ADA YONATH
Department of Structural Biology
Weizmann Institute, Rehovot 76100, Israel
Tel: 972-8-9343028
Fax: 972-8-9344154
E-mail: ada.yonath@weizmann.ac.il

Abstract: The peptidyl transferase center (PTC) is an arched void has dimensions suitable for accommodating the 3′ ends of the A-and the P-site tRNAs. It is situated within a universal sizable symmetrical region that connects all ribosomal functional centers involved in amino-acid polymerization. The linkage between the elaborate PTC architecture and the A-site tRNA position revealed that the A- to P-site passage of the tRNA 3′ end is performed by a rotatory motion, which is synchronized with the overall tRNA/mRNA sideways movement, and leads to stereochemistry suitable for peptide bond formation and for substrate mediated catalysis, thus suggesting that the PTC evolved by gene fusion. Adjacent to the PTC is the entrance of the protein exit tunnel, shown to play active roles in sequence-specific gating of nascent chains and in responding to cellular signals. This tunnel also provides a site that may be exploited for local cotranslational folding and seems to assist nascent chain trafficking into the hydrophobic space formed by the first bacterial chaperone, the trigger factor. Many antibiotics target ribosomes. Although the ribosome is highly conserved, subtle sequence and/or conformational variations enable drug selectivity, thus facilitating clinical usage. Comparisons of high-resolution structures of complexes of antibiotics bound to ribosomes from eubacteria resembling pathogens, to an archaeon that shares properties with eukaryotes and to its mutant that allows antibiotics binding, demonstrated the unambiguous difference between mere binding and therapeutical effectiveness. The observed variability in antibiotics inhibitory modes, accompanied by the elucidation of the structural basis to antibiotics mechanism justifies expectations for structural based improved properties of existing compounds as well as for the development of novel drugs.

Keywords: antibiotics selectivity; elongation arrest; resistance; ribosomal antibiotics; ribosomal symmetrical region; trigger factor.

1. Introduction

Ribosomes are giant ribonucleoprotein cellular assemblies that translate the genetic code into proteins. They are built of two subunits of unequal size that associate upon the initiation of protein biosynthesis to form a functional particle and dissociate once this process is terminated. The bacterial ribosomal subunits are of molecular weights of 0.85 and 1.45 MDa. The small

J. D. Puglisi (ed.), Structure and Biophysics – New Technologies for Current Challenges in Biology and Beyond, 127–153.
© 2007 *Springer.*

subunit (called 30S in prokaryotes) contains an RNA chain (called 16S) of about 1,500 nucleotides and 20–21 proteins, and the large one (called 50S in prokaryotes) has two RNA chains (23S and 5S RNA) of about 3,000 nucleotides in total, and 31–35 proteins. Protein biosynthesis is performed cooperatively by the two ribosomal subunits and several nonribosomal factors, assisting the fast and smooth processivity of protein formation, required for cell vitality. While elongation proceeds, the small subunit provides the decoding-center and controls translation fidelity, and the large one contains the catalytic site, called the peptidyl-transferase-center (PTC), as well as the protein exit tunnel.

mRNA carries the genetic code to the ribosome, and tRNA molecules bring the protein building block, the amino acids, to the ribosome. These L-shape molecules are built mainly of double helixes, but their two functional sites, namely the anticodon loop and the CCA 3'end, are single strands. The ribosome possesses three tRNA binding site, the A-(aminoacyl), the P-(peptidyl), and the E-(exit) sites. The tRNA anticodon loop interacts with the mRNA on the small subunit, whereas the tRNA acceptor stem, together with the aminoacylated or peptidylated tRNA 3'ends interacts with the large subunit. Hence, the tRNA molecules are the entities combining the two subunits, in addition to the intersubunit bridges, which are built of flexible components of both subunits. The elongation cycle involves decoding, the creation of a peptide bond, the detachment of the P-site tRNA from the growing polypeptide chain and the release of a deacylated tRNA molecule and the advancement of the mRNA together with the tRNA molecules from the A- to the P- and then to the E-site. This motion is driven by GTPase activity.

Two decades of experimentation (reviewed in Yonath, 2002) yielded high resolution structures of the small ribosomal subunit from *Thermus thermophilus*, T30S (Schluenzen et al., 2000; Wimberly et al., 2000), of the large subunit from the archaeon *Haloarcula marismortui*, H50S (Ban et al., 2000), from the eubacterium *Deinococcus radiodurans*, D50S (Harms et al., 2001) and recently also of the entire apo 70S ribosome (Schuwirth et al., 2005). Together with the additional structures of their complexes with substrate analogs (Bashan et al., 2003a; Hansen et al., 2002a; Nissen et al., 2000; Schmeing et al., 2002; Yusupov et al., 2001) and with a medium resolution structure of the whole ribosome from *T. thermophilus*, T70S in complex with three tRNA molecules (Yusupov et al., 2001), these structures shed light on the vast amount of biochemical knowledge accumulated in over five decades of ribosomal research.

The actual reaction of peptide bond formation is performed by a nucleophilic attack of the primary amine of the A-site amino acid on the carbonyl carbon of the peptidyl tRNA at the P-site. This reaction can be performed by tRNA 3'end analogs. Puromycin is a universal inhibitor mimicking the tip of the tRNA 3'end. Its binding to the ribosome in the presence of an active donor substrate can result in peptide bond formation uncoupled from the translocation of the A-site tRNA, namely from the polymerization of the amino acids into polypeptides. Puromycin has been commonly used as a minimal substrate for investigating the fomation of peptide bonds, in a process called the "fragment reaction", which yields a single peptide bond.

The finding that ribosomal RNA catalyzes the "fragment reaction" (Noller et al., 1992); the localization of the PTC in an environment rich in conserved nucleotides (Harms et al., 2001; Yusupov et al., 2001) the usage of puromycin derivatives bound to the partially disordered large subunits, H50S (Nissen et al., 2000), together with a compound originally presumed to resemble the reaction intermediate (Moore and Steitz, 2003), but later found to be wrongly assigned (Schmeing et al., 2005a) led to the suggestion that ribosome catalysis resembled the reverse reaction of serine proteases, and that specific ribosome nucleotides participate in the chemical events of peptide bond formation, as a "general base" (Nissen et al., 2000).

Biochemical, kinetic, and mutational results (Barta et al., 2001; Polacek et al., 2003; Sievers et al., 2004; Thompson et al., 2001; Weinger et al., 2004; Youngman et al., 2004) and the finding that the PTC conformation in crystalline H50S hardly resembles the active one (Bayfield et al., 2001), challenged this hypothesis, and indicated that there is no ground for the expectation that a complex assembly such as the ribosome catalyzes protein biosynthesis by the reverse of a common enzymatic mechanism. Indeed, the well-ordered structure of the large ribosomal subunit from *D. radiodurans*, D50S (Harms et al., 2001), determined under conditions resembling its optimal growth environment, raveled that the striking ribosomal architecture provides all structural elements enabling its function as a amino acid polymerase that ensures proper and efficient elongation of nascent protein chains in addition to the formation of the peptide bonds (Agmon et al., 2003, 2004, 2005; Baram and Yonath, 2005; Bashan and Yonath, 2005; Bashan et al., 2003a, b; Yonath, 2003a, b; 2005; Zarivach et al., 2004).

Being a prominent key player in a vital process, the ribosome is targeted by many antibiotics of diverse nature. Consequently, since the beginning of therapeutic administration of antibiotics, ribosomal drugs have been the subject to numerous biochemical and genetic studies (reviewed in Auerbach et al., 2002, 2004; Courvalin et al., 1985; Gale et al., 1981; Gaynor and Mankin, 2003; Katz and Ashley, 2005; Knowles et al., 2002; Poehlsgaard and Douthwaite, 2003; Sigmund et al., 1984; Spahn and Prescott, 1996; Vazquez, 1979; Weisblum, 1995; Yonath, 2005; Yonath and Bashan, 2004). These findings were enforced by the lessons learned from the high resolution structures of their complexes with ribosomal particles (Berisio et al., 2003a, b; Brodersen et al., 2000; Carter et al., 2000; Hansen et al., 2002b, 2003; Harms et al., 2004; Pfister et al., 2004, 2005; Pioletti et al., 2001; Schluenzen et al., 2001, 2003, 2004; Tu et al., 2005), which were found indispensable for illustrating the basic mechanisms of antibiotics activity and synergism. They also provided the structural basis for mechanisms of antibiotic resistance and enlightens the principles of antibiotics selectivity, namely the discrimination between pathogens and eukaryotes, the key for therapeutical usefulness (Auerbach et al., 2004; Yonath, 2005; Yonath and Bashan, 2004).

Since X-ray crystallography requires diffracting crystals, and since so far no ribosomes from pathogenic bacteria could be crystallized, the crystallographic studies are confined to the currently available crystals. The findings that *E. coli* and *T. Thermophilus* are practically interchangeable (Gregory et al., 2005; Thompson and Dahlberg, 2004) and that both crystallizable ribosomes are from eubacteria which resemble pathogens, permit considering them as suitable pathogen models for ribosomal antibiotics. Genetically engineered pathogen models, such as *Mycobacterium smegmatis*, can also serve as pathogen models. These should be advantageous, as they can provide isogenic mutations (Pfister et al., 2004). Similarly, for mutagenesis studies species with single rRNA operon chromosomal copy, such as *Halobacterium halobium* (Mankin and Garrett, 1991; Tan et al., 1996) are beneficial. Additional concern relates to the relevance of the crystallographic results. The ability to rationalize biochemical, functional and genetics observations by the crystallographic structures demonstrate the inherent reliability of the crystallographic results. The consistencies of drug locations with biochemical and resistance data, alongside the usage of crystalline complexes obtained at clinically relevant drug concentrations, manifest further the reliability of the crystallographic results. Last, the similarities of the structures of T30S wild type as well as of its complexes with antibiotics, elucidated by two independent laboratories (Brodersen et al., 2000; Carter et al., 2000; Pioletti et al., 2001; Schluenzen et al., 2000; Wimberly et al., 2000), indicate that dissimilarities observed crystallographically reflect genuine variability in drug binding modes.

This article focuses on the ribosomal architectural elements that govern both the positional and the chemical contributions to the catalysis of peptide bond formation, sheds light on the essentiality of accurate substrate placement and portrays the parameters dictating it; points at evolution aspects implicated by the ribosomal symmetry; describes how the first chaperon to be encountered by the nascent chain contributes to the mature protein correct folding; and also points at a possible correlation between peptide bond formation, nascent protein progression, cotranslational folding, and cellular regulation. It also relates the structural findings associated with ribosomal antibiotics action and highlights the unique achievements of these studies as well as their shortcoming. Full coverage of the vast amount of biochemical, structural, and medical knowledge is beyond the scope of this article. Instead, it emphasizes the structural finding associated with antibiotics selectivity and synergism, and describes current issues concerning to the acute problem of resistance to antibiotics.

2.   Peptide Bond Formation

2.1.   SYMMETRY WITHIN THE ASYMMETRIC RIBOSOME

The recently determined three-dimensional structures of ribosomal particles from eubacteria and archaea revealed that the interface surfaces of both subunits are rich in RNA (Figure 1), and localized the PTC in a protein-free environment the middle of the large subunit, thus confirming that the ribosome is a ribozyme. Further analysis, showed that the peptide bond is being formed within a universal sizable symmetrical region (Figure 2), containing ~180 nucleotides (Agmon et al., 2003, 2004, 2005; Baram and Yonath, 2005; Bashan and Yonath, 2005; Bashan et al., 2003a, b; Yonath, 2003a, b, 2005; Zarivach et al., 2004). The symmetrical region is located in and around the PTC, and its symmetrical axis, which is directed into the protein exit tunnel, passes through the peptidyl transferase center, midway between the RNA features shown to host the 3'ends of the A- and the P- sites tRNA.

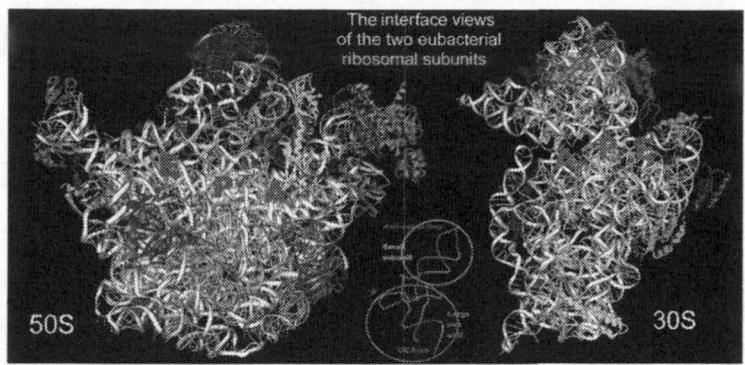

Figure 1. The two ribosomal subunits. The small (30S) and the large (50S) ribosomal subunit, from *T. thermophilus* (Schluenzen et al., 2000) and *D. radiodurans* (Harms et al., 2001), respectively, showing their intersubunit interfaces. In both, the ribosomal RNA is shown as *silver ribbons*, and the ribosomal proteins main chains in different colors. A, P, E designate the approximate locations of the A-,P-, and E- tRNA anticodons on the small subunit, and the tRNA beginning of the tRNA acceptor stems (*the red star on the inserted figure*) on the large one. The regions of tRNA interactions with each subunit are shown on the tRNA molecule, *inserted in the middle*. The *red star* indicates the position at which the tRNA acceptor stem meets the large subunit.

Although first identified in D50S, this symmetrical region seems to be a universal ribosomal feature, as it is present in all known structures of the large ribosomal subunit (Figure 2). The symmetrical region extends far beyond the vicinity of the peptide synthesis location and interacts, directly or through its extensions, with all ribosomal functional features that are relevant to the elongation process: the tRNA entrance and exit regions, namely the L7/L12 stalk and the L1 arm, respectively, the peptidyl transferase center, and the bridges connecting the two subunits (Figure 2), among which bridge B2a resides on the PTC cavity and reaches the vicinity of the decoding center in the small subunit (Yusupov et al., 2001). The 3′ends of the A- and the P- tRNAs bind to the PTC, and even the 3′ end of the E-site tRNA contacts the neighborhood of the symmetry region edge in the T70S complex (Yusupov et al., 2001) but not in H50S complexed with a fragment of the E-site tRNA (Schmeing et al., 2003). Hence, the spatial organization of this region and its central location may enable signal transmission between the remote locations on the ribosome (Agmon et al., 2003).

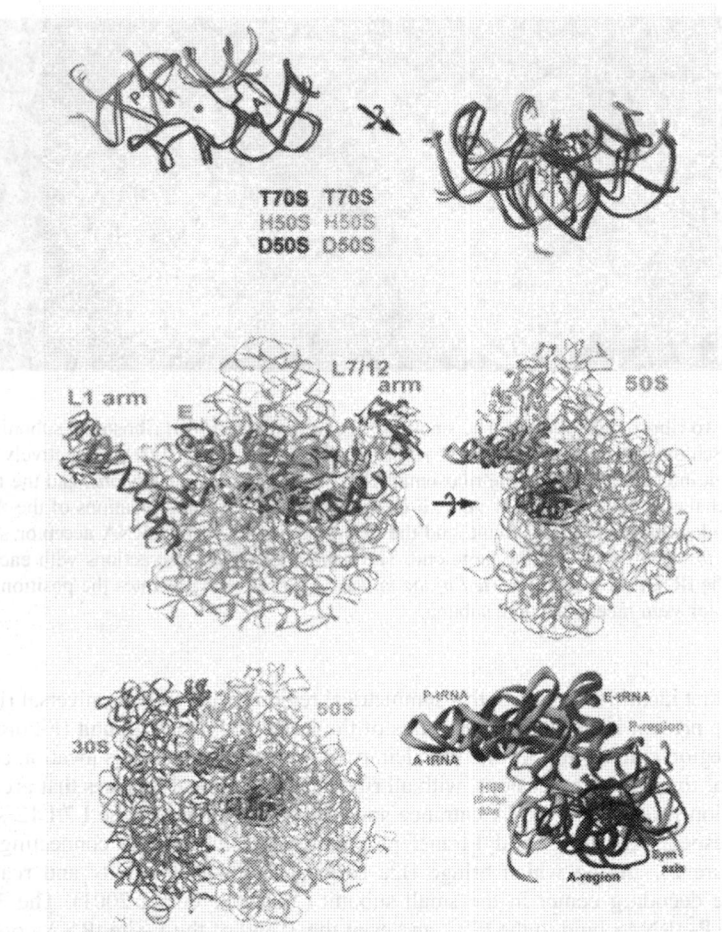

Figure 2. The symmetrical region within the large ribosomal subunit.Throughout, the part containing the A-loop (namely the site of A-site tRNA 3′end) is *blue* (called: *A-region*), and the corresponding one, containing the P-site tRNA 3′end, is *green*. Similarly, the A-site tRNA mimic is shown in *blue* and the derived P site tRNA is *green*. The symmetry axis is shown in *red*. (a) Two orthogonal views (*top* and *side, respectively*) of the superposition of the backbone of the symmetrical regions in all known structures: the entire ribosome from T70S (PDB 1GIY), D50S (PDB 1NKW), and H50S (PDB 1JJ2). Note that the A-site mimic and the derived P-site are incorporated into the side view. *Middle and bottom*: The symmetry related region within the large subunit (*upper panel*) and the entire ribosome (*bottom left*). The direct extensions of the symmetrical region are shown in *purple*. Ribosomal RNA in shown is *gray*

2.2. THE RIBOSOME IS A POLYMERASE

Located at the bottom of a V-shaped cavity (Figure 3), the PTC is an arched void with dimensions suitable for accommodating the 3'ends of the A- and the P-site tRNAs. Each of the symmetry related subregions contains half of the PTC, namely either the A- or the P-site, and the axis relating them by ~180° rotation, is located in the middle of the PTC, midway between the two tRNA binding sites. In a complex of D50S with a 35-nucleotides oligomers mimicking the aminoacylated-tRNA acceptor stem, called ASM (Figure 3), the bond connecting the 3'end with the acceptor stem was found to roughly coincide with the symmetry axis (Bashan et al., 2003a), suggesting that tRNA A → P-site passage is a combination of two independent, albeit synchronized motions: a sideways shift of most of the tRNA molecules, performed as a part of the overall mRNA/tRNA translocation, and a rotatory motion of the tRNA 3'end within the PTC. The path provided by the rotatory motion is confined by the PTC rear wall and by two nucleotides that bulge from the front wall into the PTC center.

Simulation of the rotatory motion (Figure 3) revealed that it is navigated and guided by striking architectural design of the PTC, and that it terminates in a stereochemistry appropriate for a nucleophilic attack of the A-site amino acid on the carbonyl carbon of the peptidyl tRNA at the P-site (Agmon et al., 2003, 2005; Bashan et al., 2003a, b). The spatial match between the PTC rear wall and the contour of the tRNA aa-3'end, formed by the rotatory motion, indicates that it provides the template for the translocation path. From the other side of the PTC, two universally conserved nucleotides A2602 and U2585 (*E. coli* nomenclature, throughout), bulge towards the PTC center (Figures 3) and seem to anchor and/or propel the rotatory motion (Agmon et al., 2003, 2004, 2005; Baram and Yonath, 2005; Bashan et al., 2003a, b; Polacek et al., 2003; Zarivach et al., 2004).

Importantly, the derived P-site tRNA 3'end forms all interactions found biochemically (e.g. Bocchetta et al., 1998; Green et al., 1997) and the orientation of the so created peptide bond is adequate for the ribosomal subsequent tasks, including the release of the peptidyl-tRNA and the entrance of the nascent protein into the exit tunnel. Hence, it appears that the ribosome provides a striking architectural frame, ideal for amino acid polymerization. Thus, the ribosome functions as an enzyme, a ribozyme, responsible not only to peptide bond formation, but also for the successive reactions, namely the creation of polypeptides that can eventually acquire their functional fold (Agmon et al., 2003, 2004, 2005; Bashan et al., 2003a; Zarivach et al., 2004).

---

ribbons. The positions of the docked three tRNA molecules, as seen in the complex of T70S (PDB 1GIY) are also shown, to indicate their relationship to the symmetry related area. The *gold* feature is the intersubunit bridge (B2a) that combines the two ribosomal active sites. An enlarged view of the symmetry-related region is shown in *right bottom corner*. Note the strategic location of H69, which bridges the two subunits, and plays a major role in A-site tRNA accurate placement.

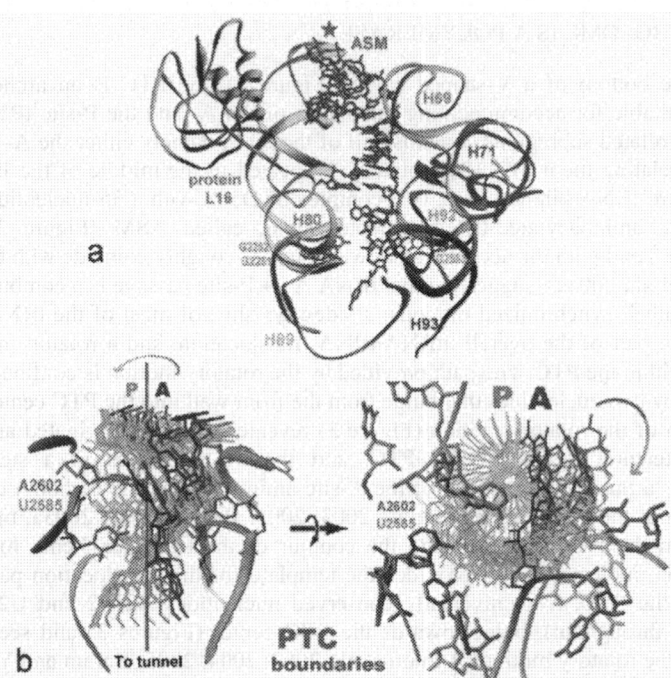

Figure 3. The rotatory motion. Throughout, the part containing the A-loop (namely the site of A-site tRNA 3′end) is *blue* (called: *A-region*), and the corresponding one, containing the P-site tRNA 3′end, is *green*. Similarly, the A-site tRNA mimic is shown in *blue* and the derived P-site tRNA is *green*. The symmetry axis is shown in *red*. (a) The PTC pocket, including ASM, an A-site substrate analog, which is represented by atoms in *red*. The *red star* indicates the position at which the tRNA acceptor stem meets the large subunit, as in Figure 1. The RNA components of the PTC pocket are numbered according to *E. coli* nomenclature (also shown in Figure 5) and colored differently. Note the remote interactions positioning the substrate, as well as the universal contributors to the 3′end base pairs (*a single basepair at the A-site, and two in the P-site*). (b) Two orthogonal snapshots (*sideways and from the tunnel into the PTC*) of intermediate stages (*represented by gradual transformation from blue to green*) in the motion of the A-site tRNA CCA from the A- to the P-site. The two front-wall bulged nucleotides are shown in *pink* and *magenta*. The simulation was performed by rotating the ASM aminoacylated 3′end by (10 times 18° each) within D50S PTC, around the bond connecting the ASM 3′end with its acceptor stem, accompanied by a 2 Å shift in the direction of the tunnel, as implied by the overall spiral nature of the PTC template. The *blue-green round arrows* show the rotation direction. The ribosomal components belonging to the PTC rear wall, that confine the exact path of the rotatory motion, are shown in *gold*. The two front wall flexible nucleotides, A2602 and U2585, are colored in *magenta* and *pink*, respectively.

2.3. SUBSTRATE POSITIONING AND PTC TOLERANCE

Positioning reactants in orientation suitable for chemical reactions is performed by almost all biocatalysts (Jencks, 1969, reissued 1987). Different from enzymes catalyzing a single chemical reactions, such as proteases, and similar to other polymerases, the ribosome provides the means not only for the chemical reaction (peptide bond formation), but also for substrates motions required for the processivity of peptide bond formation, namely for amino acid polymerization. However, a prerequisite for achieving the ribosome contribution is accurate substrate placement (Yonath, 2003a, b).

The universal Watson–Crick base pair between C75 of A- site tRNA terminus and G2553 (Figure 3) (Kim and Green, 1999), and its symmetrical mate at the P-site, namely the base pair between C75 and G2251, exist in all known structures (Bashan et al., 2003a; Hansen et al., 2002a; Nissen et al., 2000; Schmeing et al., 2002; Yusupov et al., 2001). Positioning governed solely by this base pair is sufficient for entropy driven peptide -bond formation (Gregory and Dahlberg, 2004; Sievers et al., 2004). However, it may not suffice for allowing smooth amino acid polymerization, as shown by the correlation found between the rates of peptide bond formation and the substrate type. Thus, compared to the reaction rate with full size tRNAs, when using the minimal substrate puromycin, such as the "fragment reaction" reactants, the peptide bond is being formed at significantly reduced rates (Moore and Steitz, 2003). Consistently, the locations and orientations of all "fragment reaction" reactants in ribosomal crystals indicate a need to undergo repositioning and/or rearrangements in order to participate in peptide bond formation (Moore and Steitz, 2003; Yonath, 2003b). This time-consuming process can be responsible to the slowness of the "fragment reaction". These observations indicate that the A-site base pairing is not sufficient for accurate tRNA placement, essential for performing the rotatory motion. As the main structural difference between fragment reaction reactants and full-size tRNA is the substrates relative sizes, it appears that accurate positioning is achieved by remote interactions of the A-site tRNA acceptor stem with the upper part of the PTC cavity (Agmon et al., 2003; Yonath, 2003a, b).

Remote interactions cannot be formed by substrate analogs that are too short to reach the PTC cavity upper part, as "fragment reaction" participants, or when helix H69, the remote interactions mate at the PTC upper end (Figures 2 and 3) is disordered, as in H50S structure (Ban et al., 2000; Nissen et al., 2000). It appears, therefore, that the CCA base pairing contributes to the overall positioning of the 3'end of the aminoacylated tRNA, whereas the efficiency of peptide bond formation depends on the tRNA remote interactions. The rotatory motion guides the A-site tRNA to land at the P-site in an orientation appropriate for the creation of the two basepairs. This double basepair seem to stabilize the orientation of P-site tRNA at the conformation essential for the P-site tRNA catalytic role in peptide bond formation (Dorner et al., 2002; Weinger et al., 2004). Hence, the rotatory motion not only leads to a configuration suitable for peptide bond formation (Agmon et al., 2003; Bashan et al., 2003a), it also places the reactants at a distance reachable by the O2' of the P- site tRNA A76.

Remote placement of the A-site 3'end of the tRNA seems to be designed to tolerate variability in PTC binding, as it is required to comply with the ability of the ribosome to accommodate all of the amino acids, to allow for the rotatory motion, and to undergo induced fit of compounds that mimic the real substrate only partially (Schmeing,et al., 2005b). It appears therefore that the tRNA size and shape and the overall ribosome architecture determines the position of the tRNA molecules and the universal base pairs, described above, establish the approximate inclination of the A-site tRNA 3'end, and facilitates P-site meditated catalysis. Accurate A-site tRNA alignment, however, is governed by its remote interactions, and since such placement is the prerequisite for the processivity of protein biosynthesis, it appears that the role played by the remote interactions supersedes all others. This conclusion is supported by the finding that in the absence of these interactions, similar, albeit distinctly different, binding modes are formed, which contrary to substrate orientation dictated by remote interactions, leads to optimal stereochemistry for the formation of a peptide bond. Hence, binding independent of remote directionality leads to various orientations, each requiring conformational rearrangements to participate in formation of a peptide bond (Moore and Steitz, 2003).

In short, by identifying the linkage between the universal ribosomal symmetry and the substrate binding mode, the integrated ribosomal machinery for peptide bond formation, amino acid polymerization, and translocation within the PTC, was revealed (Agmon et al., 2003; Bashan et al., 2003a). This machinery is consistent with results of biochemical and kinetic studies (Gregory and Dahlberg, 2004; Nierhaus et al., 1980; Sievers et al., 2004; Youngman et al., 2004), proposing that positioning of the reactive groups is the critical factor for the catalysis of intact tRNA substrates, and does not exclude assistance from ribosomal or substrate moieties. Hence, by offering the frame for correct substrate positioning, as well as for catalytic contribution of the P-site tRNA 2'-hydroxyl group, as suggested previously (Dorner et al., 2002; Weinger et al., 2004), the ribosomal architectural-frame governs the positional requirements, and provides the means for substrate mediated chemical catalysis.

2.4. PTC MOBILITY AND ANTIBIOTICS SYNERGISM

The two universally conserved nucleotides A2602 and U2585 that bulge towards the PTC center (Figures 3B and C) and do not obey the symmetry, are extremely flexible. In D50S A2602 is placed beneath A73 of A-site tRNA, within contact distance throughout the course of the rotation. Similarly, U2585, situated under A2602 and closer to the tunnel entrance, is located within a contact distance to bound amino acid throughout the A- to P-site motion. Nucleotide A2602 exhibits a large variety of conformations in different complexes of the large subunit (Agmon et al., 2003; Bashan et al., 2003a). A2602 is involved in several tasks other than peptide bond formation, such as nascent peptide release (Polacek et al., 2003) and anchoring tRNA A- to P-site passage (Agmon et al., 2003, 2005; Bashan et al., 2003a, b; Zarivach et al., 2004).

Sparsomycin, which target A2602 (Bashan et al., 2003a; Hansen et al., 2003; Porse et al., 1999), is a potent universal antibiotics agent, hence less useful as anti-infective drug. Comparisons between sparsomycin binding sites in D50S (Bashan et al., 2003a) and H50S (Hansen et al., 2003) indicated the correlation between antibiotics binding mode and the ribosomal functional-state. By binding to non-occupied large ribosomal subunits, sparso-mycin stacks to A2602 and causes striking conformational alterations in the entire PTC, which should influence the positioning of the tRNA in the A-site, thus explaining why sparso-mycin was considered to be an A-site inhibitor, although it does not interfere with A-site substrates (Goldberg and Mitsugi, 1966; Monro et al., 1969; Porse et al., 1999). Within D50S, sparsomycin faces the P-site. Hence, it can also enhance nonproductive tRNA-binding (Monro et al., 1969). Conversely, when sparsomycin enters the large subunit simultaneously with a P-site substrate or substrate-analog, it can cause only a modest conformation alteration of A2602, and because the P-site is occupied by the P-site substrate, sparsomycin stacking to A2602 appears to face the A-site (Hansen et al., 2003).

The base of U2585 undergoes a substantial conformational alteration in a complex of D50S with Synercid – a synergetic antibiotic agent, of which one part binds to the PTC and the other blocks the protein exit tunnel (Agmon et al., 2004; Harms et al., 2004). This recently approved injectable drug with excellent synergistic activity, is a member of the streptogramins antimicrobial drug family in which each drug consists of two synergistic components (SA and SB), capable of cooperative converting weak bacteriostatic effects into lethal bactericidal activity.

In crystals of D50S-Synercid complex, obtained at clinically relevant concentrations, the SA component, dalfopristin, binds to the PTC and induces remarkable conformational alterations; including a flip of 180° of U2585 base hence paralyze its ability to anchor the rotatory motion and to direct the nascent protein into the exit tunnel (Agmon et al., 2004). As the motions of U2585 are of utmost importance to cell vitality, it is likely that the pressure for maintaining the processivity of protein biosynthesis will attempt recovering the correct positioning of U2585, by expelling or relocating dalfopristin, consistent with dalfopristin low antibacterial effect. The SB component of Synercid, quinupristin, is a macrolide that binds to the common macrolide-binding pocket (Auerbach et al., 2004; Schluenzen et al., 2001). Due to its bulkiness, quinupristin is slightly inclined within the tunnel, and consequently does not block it efficiently (Agmon et al., 2004; Harms et al., 2004), thus rationalizing its reduced antibacterial effects compared to erythromycin.

Since within the large ribosomal subunit both Synercid components interact with each other, the nonproductive flipped positioning of U2585 is stabilized, and the way out of dalfopristin is blocked. Hence, the antimicrobial activity of Synercid is greatly enhanced. Thus, the two components of this synergetic drug act in two radically different fashions. Quinupristin, the SB component, takes a passive role in blocking the tunnel, whereas dalfopristin, the SA component, plays a more dynamic role by hindering the motion of a vital nucleotide at the active site, U2585. It is conceivable that such mode of action consumes higher amounts of material, compared to the static tunnel blockage, explaining the peculiar composition of 7:3 dalfopristin/quinupristin in the optimized commercial Synercid, although the crystal structure of the complex D50S-Synercid indicates binding of stoichiometric amounts of both components.

The mild streptogramins reaction on eukaryotes may be linked to the disparity between the 180° flip of U2585 in D50S (Harms et al., 2004) and the mild conformational alterations of U2585 imposed by the SA compounds on eukaryotic or archaeal ribosome, as seen in the complex of H50S with Virginiamycin-M, a streptograminA component (Hansen et al., 2002b). This significant difference in binding modes to eubacterial vs. archaeal ribosomes appears to reflect the structural diversity of PTC conformations (Harms et al., 2001; Yonath, 2002; Yusupov et al., 2001), consistent with the inability of H50S PTC to bind the peptide bond formation blocker clindamycin, as well as the A-site tRNA competitor chloramphenicol (Mankin and Garrett, 1991).

3. On Ribosome Evolution

The entire symmetrical region is highly conserved, consistent with its vital function. Sampling 930 different species from three phylogenetic domains (Cannone et al., 2002) shows that 36% of all of E. coli 23S RNA nucleotides, excluding the symmetrical region, are "frequent" (namely, found in > 95% of the sequences), whereas 98% of the symmetrical region nucleotides are categorized as such. The level of conservation increases in the innermost shell of the symmetry related region. Thus, among the 27 nucleotides lying within 10 Å distance from the symmetry axis, 75% are highly conserved, among these seven are absolutely conserved.

The universality of the symmetrical region hints that the ribosomal active site evolved by gene fusion of two separate domains of similar structures, each hosting half of the catalytic activity. Importantly, whereas the ribosomal internal symmetry relates nucleotide orientations and RNA backbone fold (Figure 2), there is no sequence identity between related nucleotides in the A- and the P-regions. The preservation of the three-dimensional structure of the two halves of the ribosomal frame regardless of the sequence demonstrates the rigorous requirements of accurate substrate positioning in stereochemistry supporting peptide bond formation.

Similarly, protein L16, the only ribosomal protein contributing to tRNA positioning (Agmon et al., 2003; Bashan et al., 2003a), displays conserved tertiary structure alongside diverged primary sequence. Consistently, results of recent experiments addressing the functional conservations of the ribosome, show that the translational factor function and subunit–subunit interactions are conserved in two phylogenetically distant species, E. coli and T. thermophilus, despite the extreme and highly divergent environments to which these species have adapted (Thompson and Dahlberg, 2004). Similarly, mutations in T. thermophilus 16S and 23S rRNAs, within the decoding site and the PTC, produced phenotypes that are largely identical to their mates in mesophilic organisms (Gregory et al., 2005).

The contribution of protein L2 to the ribosomal polymerase activity may also shed some light on ribosome evolution. Protein L2 is the only protein interacting with both the A- and the P-regions (Agmon et al., 2005), and between its two residues involved in these interactions, one (229) was shown to be essential for the elongation of the nascent chain (Cooperman et al., 1995). It appears, therefore, that the main function of L2 is to provide stabilization to the PTC while elongation takes place. Stabilization of the ribosomal frame is mandatory for maintaining accurate substrate positioning, which, in turn, is required for enabling the rotatory motion, but is irrelevant to single peptide bond formation (Yonath, 2003b). This finding is consistent with the assumption that the ancient ribosome was made only from RNA and that the proteins were added later, in order to increase its fidelity and efficiency.

Involvement in maintaining the symmetry region architecture, and consequently in peptidyl transferase activity can also be attributed to protein L36. This small Zn containing protein is situated in the middle of four parallel helixes and seems to stabilize their overall conformation. Two of these helixes are part of the symmetry related region and two are the nonsymmetrical extensions of the PTC main components. Furthermore, at its location, L36 interactions can also connect these helixes with the elongation factors binding sites. Hence, in addition to stabilizing the conformation of the symmetry related region, it may also be involved in transmitting information about factor binding. The possible availability of alternative route for signaling and/or alternative means for conformation preservation, may account for the absence of L36 in some species, such as H. marismortui.

4.   The Ribosomal Tunnel

4.1.   ELONGATION ARREST AND TUNNEL MOBILITY

Nascent proteins emerge out of the ribosome through an exit tunnel, a universal feature of the large ribosomal subunit first seen in the mid-eighties (Milligan and Unwin, 1986; Yonath et al., 1987). This tunnel is adjacent to the PTC and its opening is located at the other end of the subunit (Figure 4). Lined primarily by ribosomal RNA, this tunnel is rather kinked, has a nonuniform diameter, and contains grooves and cavities (Ban et al., 2000; Harms et al., 2001). Among the few r-proteins reaching its wall, the tips of extended loops of proteins L4 and L22 create an internal constriction.

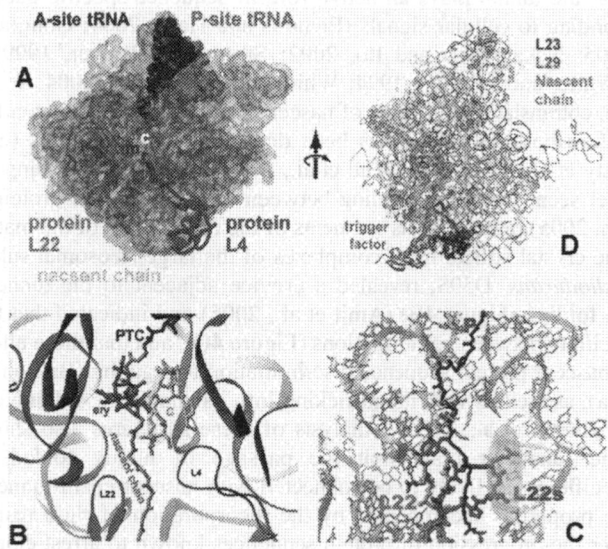

Figure 4. The ribosome tunnel. (**A**) A section through D50S (*in light purple*) with docked A- and P-sites tRNA and modeled polyalanine nascent chain (*yellow*). The approximate positions of the PTC (*P*), the hydrophobic crevice (*C*), and the macrolide pocket (*m*) are marked. The main chains of proteins L4 and L22 are shown in *red* and *cyan*, respectively. (**B**) The hydrophobic crevice (*C*), in relation to the PTC, the macrolide-binding site, represented by erythromycin (ERY), to the tunnel constriction composed of the tips of the elongated loops of proteins L4 and L22, and to the possible pat of the nascent chain (*modeled as polyalanine and shown in green*), Rapamycin binding mode is shown in gold. (**C**) A view parallel to the tunnel long axis (*rRNA in olive green*) with a modeled nascent chain (*blue*). The tip of the ribosomal protein L22 beta-hairpin at its native and swung (L22S) conformations, the latter induced by troleandomycin (*T, in gold*) binding, are shown in *cyan* and *magenta*, respectively. The modeled polypeptide chain (*blue*) represents a nascent protein with the sequence motif known to cause SecM (secretion monitor) elongation arrest. This motif is located about 150 residues from the N-terminus and has the sequence XXXWXXXXXXXXXXP, where X is any amino acid and P (proline) is the last amino acid to be incorporated into the nascent chain (based on Nakatogawa and Ito, 2002). The positions of two key residues for nascent protein arrest, proline and tryptophane, are highlighted in *red*, to indicate the stunning correlation between its position and that of troleandomycin (*T, in gold*). The specific proline of SecM that is required for the arrest when incorporated into the protein at the PTC is the top amino acid of the modeled nascent chain is designated by P. The shaded area designates the region where mutants bypassing the arrest were depicted (Nakatogawa and Ito, 2002). (**D**) A side view of the structure of trigger factor in complex with D50S (*represented by purple-brown RNA backbone and purple-pink ribosomal proteins*). The bound trigger factor binding domain is shown in *orange*, and a modeled polypeptide chain in *cyan*. Ribosomal proteins L29 and L23 are highlighted in *magenta* and *blue*, respectively. Note the elongated loop of L23, a unique eubacterial feature, which reaches the interior of the tunnel, to a location allowing its interaction with the emerging nascent chain.

Five years ago, when first observed at high resolution in H50S crystal structure, this tunnel was assumed to be a firmly built passive and inert conduit for nascent chains (Ban et al., 2000; Nissen et al., 2000). However, biochemical results, accumulated during the last decade, indicate that the tunnel plays an active role in sequence-specific gating of nascent chains and in responding to cellular signals (Etchells and Hartl, 2004; Gong and Yanofsky, 2002; Johnson, 2005; Nakatogawa and Ito, 2002; Stroud and Walter, 1999; Tenson and Ehrenberg, 2002; Walter and Johnson, 1994; White and von Heijne, 2004; Woolhead et al., 2004). Furthermore, cotranslational folding of nascent polypeptides into secondary structures while still within the ribosomal tunnel has been detected in several cases (e.g. Eisenstein et al., 1994; Hardesty et al., 1995; Woolhead et al., 2004). Such initial folding events within the ribosomal tunnel seem to serve signaling between the cell and the protein-biosynthetic machinery (Johnson, 2005) rather than as segments of the correct fold of the mature protein.

Consistently, the crystal structures of complexes of the large ribosomal subunit from the eubacterium *D. radiodurans*, D50S, revealed a crevice adjacent to the tunnel that can be exploited for initial folding (Figure 4B) (Amit et al., 2005) and indicated that the tunnel has the capability to oscillate between conformations (Figure 4C), and that these alterations could be correlated with nascent protein sequence discrimination and gating (Bashan et al., 2003b; Berisio et al., 2003a), as well as with its trafficking into its chaperone-folding cradle (Baram and Yonath, 2005; Baram et al., 2005). Analysis of these structures also shows that at its entrance, the tunnel diameter may limit the passage of highly folded polypeptides. Furthermore, in specific cases, likely to be connected with nascent chain-tunnel interactions, the tunnel entrance properties accompanied by the incorporation of rigid residues, such as proline, may hamper the progression of protein sequences known to arrest elongation (Gong and Yanofsky, 2002; Nakatogawa and Ito, 2002).

So far most of the tunnel functional roles have been attributed to mobile extended loops of ribosomal proteins that penetrate its walls, which are primarily made of ribosomal RNA. Examples are the tips of extended loops of proteins L22 and L23 that seem to provide communication routes for signaling between the ribosome and the cell, as their other ends are located on the solvent side of the ribosome, in the proximity of the tunnel opening (Baram and Yonath, 2005; Berisio et al., 2003a; Harms et al., 2001). Furthermore, the beta-hairpin tip of L22 can swing across the tunnel around its accurately placed hinge (Figure 4C), and gate the tunnel. This motion appears to provide a general mechanism for elongation arrests triggered by specific cellular conditions, since the interacting nucleotides with the swung L22 hairpin tip are identical to those identified in mutations bypassing tunnel arrest (Agmon et al., 2003; Bashan et al., 2003b; Berisio et al., 2003a). Thus, this elongated ribosomal protein (Harms et al., 2001; Unge et al., 1998) that stretches along the large subunit may not only contribute to the dynamics associated with tunnel arrest, but also participate in signal transmission between the cell and the ribosomal interior.

4.2. INTRA-RIBOSOME CHAPERON ACTIVITY?

The crystal structure is a complex of the large ribosomal subunit from *D. radiodurans*, co-crystallized with rapamycin, a polyketide with no inhibitory activity, revealed that rapamycin binds to a crevice located at the boundaries of the nascent protein exit tunnel, opposite to the macrolide pocket (see below and Figures 4A and B). Being adjacent to the ribosome tunnel, but not obstructing the path of nascent chains at extended conformation, this crevice may provide the site for local cotranslational folding of nascent chains (Amit et al., 2005). The size of this crevice is suitable for accommodating small secondary structural elements, and therefore may provide nascent chains a site for adopting a particular fold at the early stage of tunnel passage, consistent with a large range of biochemical evidence, obtained mainly for transmembrane proteins, implicating cotranslational folding (Etchells and Hartl, 2004; Johnson, 2005; Stroud and Walter, 1999; Tenson and Ehrenberg, 2002; Walter and Johnson, 1994; White and von Heijne, 2004; Woolhead et al., 2004).

Similar to rapamycin, transmembrane protein segments are highly hydrophobic, and therefore may be accommodated within the crevice. Hence, this crevice may provide the space as well as the hydrophobic patch that might act as an inner-tunnel chaperone, consistent with findings interpreted as nascent chain folding near the PTC, which was proposed to correlate with sequential closing and opening of the translocon at the ER membrane (Woolhead et al., 2004). Hence the detection of the crevice confirms that the tunnel possesses specific binding properties, and suggests that this crevice plays a role in regulating nascent protein progression, thus acting as an intra-ribosome chaperon.

The cotranslational folding may be only transient, until messages are transmitted to other cell components (e.g. the translocon pore) (Etchells and Hartl, 2004; Woolhead et al., 2004). Alternatively, it is conceivable that once small nucleation centers are formed, they may progress through the tunnel by temporary expansions of the tunnel diameter, as observed recently for translation-arrested ribosomes (Gilbert et al., 2004). Cotranslational folding is frequently observed for eukaryotic membrane proteins. These may possess a comparable crevice, as a similar feature could be identified also in the archaeal H50S. However, although existence of a crevice is postulated in ribosomes from all kingdoms of life, this does not imply structural identity, since phylogenetic diversity should play a considerable role in its detailed structure, as found at the macrolide-binding pocket (Auerbach et al., 2004; Baram and Yonath, 2005; Pfister et al., 2004; Yonath, 2005; Yonath and Bashan, 2004). Hence, the binding affinities of this crevice should vary, explaining why rapamycin is not known to strongly inhibit membrane proteins translation.

4.3. THE FIRST ENCOUNTER WITH RIBOSOME ASSOCIATED CHAPERONE

The complex process of folding newly synthesized proteins into their native three-dimensional structure is vital in all kingdoms of life. Although, in principle, protein can fold with no assistance of additional factors, since their sequences entail their unique folds, under cellular conditions nascent polypeptides emerging out of the ribosomal tunnel are prone to aggregation and degradation, and thus require assistance. The cellular strategy to promote correct folding and prevent misfolding involves a large arsenal of molecular chaperones (Bukau et al., 2000; Frydman, 2001; Gottesman and Hendrickson, 2000; Hartl and Hayer-Hartl, 2002; Rospert, 2004; Thirumalai and Lorimer, 2001). These proteins are found in all kingdoms and the existence of ribosome-associated chaperones is a highly conserved principle in eukaryotes and prokaryotes, although the involved components differ between species.

In eubacteria, the folding of cytosol proteins is coordinated by three chaperone systems: the ribosome-associated trigger factor, DnaK, and GroEL. trigger factor (TF), a unique feature of eubacteria, is the first chaperone encountering the emerging nascent chain. This 48 kDa modular protein is composed of three domains, among which the TF N-terminal domain (TFa) that contains a conserved "signature motif", mediates the association with the ribosome

(Maier et al., 2005). It cooperates with the DnaK system, and their combined depletion causes a massive aggregation of newly synthesized polypeptides as well as cell death above 30°C (Deuerling et al., 1999). Biochemical studies showed that TF binds to the large ribosomal subunit at 1:1 stoichiometry by interacting with ribosomal proteins L23 and L29 (Blaha et al., 2003; Kramer et al., 2002).

Protein L23 exists in ribosomes from all kingdoms of life, but belongs to the small group of ribosomal proteins that display a significant divergence from conservation. Thus, in all species, it is built of an almost identical globular domain. In eubacteria, however, it possesses a unique feature, a sizable elongated loop, which in *D. radiodurans* extends from the vicinity of the tunnel opening all the way into the tunnel interior (Figure 4D) and in (Harms et al., 2001), and can actively interact with the nascent protein passing through it (Baram and Yonath, 2005; Baram et al., 2005), implying a possible dynamic control.

The high resolution crystals structure of D50S in complex with the TFa domain from the same source showed that the "signature motif" and its few amino acids extension (called the "extended signature motif"), interact with the large ribosomal subunit near the tunnel opening (Figure 4D) at a triple junction between ribosomal proteins L23 and L29 and the 23S rRNA (Baram et al., 2005), consistent with a previous suggestion (Kristensen and Gajhede, 2003).

Despite the similarity between the overall structures of the ribosome-bound and the unbound TFa (Ferbitz et al., 2004; Ludlam et al., 2004), significant differences were detected between their conformations and that of the bound TFa, indicating a substantial conformational rearrangement of TFa upon binding to the ribosome. These alterations result in the exposure of a sizable hydrophobic patch facing the interior of the ribosomal exit tunnel, which should increase the tunnel's affinity for hydrophobic segments of the emerging nascent polypeptide. Thus, the trigger factor prevents aggregation of the emerging nascent chains by providing a competing hydrophobic environment (Baram et al., 2005).

In D50S, protein L23 exposes a sticky hydrophobic patch, located in the wall of the ribosomal tunnel and available for interactions with hydrophobic regions of the progressing nascent chain. These interactions may be involved in cotranslational folding of nascent polypeptides into secondary structures while still within the ribosomal tunnel, and such events may trigger signaling to the cell, for recruiting TF and initiating its binding. Thus, the subjection of L23 elongated loop may affect, in turn, its interaction with TF. Similar to the undetected conformational changes in the chimeric complex (Ferbitz et al., 2004), mainly due to the disorder of the corresponding TFa region, the possible involvement of L23 loop in initial folding and/or FT attraction could not be seen in the chimeric complex, since, like in eukaryotes, L23 of H50S lacks the elongate loop that penetrates the tunnel.

It seems, therefore, that protein L23 plays multiple roles in eubacteria. It is essential for the association of TF with the ribosome, and since the tip of its internal loop can undergo allosteric conformational changes thus modulating the shape and the size of the tunnel (Baram and Yonath, 2005; Baram et al., 2005), it may control the pace of the entrance of the nascent chain into its shelter and serve as a channel for cellular communication with the nascent chain while progressing in the tunnel.

5.   Antibiotics Targeting the Ribosomal Tunnel

5.1.   ANTIBIOTICS SELECTIVITY: THE KEY FOR THERAPEUTIC EFFECTIVENESS

Ribosomes show a high level of universality in sequence and almost complete identity in function, therefore the imperative distinction between pathogens and human, the key for antibiotics usefulness, is achieved by subtle structural difference within the antibiotics binding pockets of the prokaryotic and eukaryotic ribosomes (Auerbach et al., 2004; Yonath and Bashan, 2004). Both *D. radiodurans* and *H. marismortui* are nonpathogenic organisms. Nevertheless, there are major differences between the suitability of their ribosomes to serve as pathogen models. Thus, although *D. radiodurans* is an extremely robust gram-positive eubacterium that can survive in harsh environments, it is best grown under conditions almost identical to those allowing for optimal biological activity of *E. coli* (Harms et al., 2001) and shows striking sequence similarity to it. Moreover, contrary to archaeal and halophilic ribosomes, which possess typical eukaryotic elements at the principal antibiotics targets and are not inhibited by antibiotics at the clinically useful concentrations (Mankin and Garrett, 1991; Sanz et al., 1993), *D. radiodurans* ribosomes are targeted by the common ribosomal antibiotics at clinically relevant concentrations in a fashion similar to most pathogens (Auerbach et al., 2004; Schluenzen et al., 2001). Thus, the availability of structures of antibiotics complexed with ribosomes from both species provides unique tools for investigating the structural basis for antibiotics selectivity.

A striking example is the immense influence of the minute difference between adenine and guanine in position 2058, which was found to dictate the affinity of macrolides binding. Macrolides are natural and semisynthetic compounds, which rank highest in clinical usage. They are characterized by a macrolactone ring to which at least one sugar moiety is attached (Figure 5). The first widely used macrolide drug is erythromycin, a 14-member lactone ring, decorated by a desosamine and cladinose sugars. Ketolides belong to a novel class of the macrolide family, characterized by a keto group at position 3 of the macrolactone ring, a single aminosugar moiety, and an extended hydrophobic arm (Figure 5).

This recently developed drug family was designed to act against several macrolide resistant bacterial strains. Both macrolides and ketolides were shown, crystallographically, to bind to a specific pocket in the eubacterial tunnel, called below the "macrolide-binding-pocket". Both act by producing a steric blockage of the ribosome exit tunnel, hence hampering the progression of nascent chains (Auerbach et al., 2004; Berisio et al., 2003a, b; Hansen et al., 2002b; Pfister et al., 2004, 2005; Schluenzen et al., 2001, 2003; Tu et al., 2005; Yonath, 2005; Yonath and Bashan, 2004).

This high affinity pocket is composed of nucleotides belonging to the 23S RNA (Figure 5) and is located at the upper end of the tunnel, below the PTC and above the tunnel constriction (Figure 4A). All currently available crystal structures of complexes of 14-membered ring

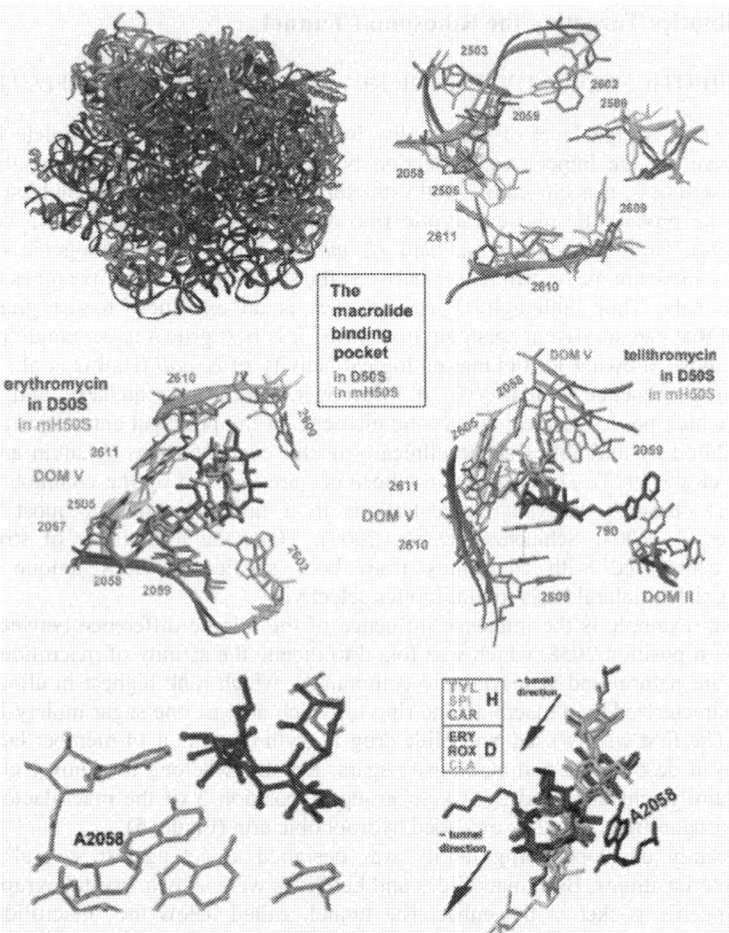

Figure 5. The macrolide-binding pocket. *Top left*: a view into D50S ribosome tunnel, with bound erythromycin (*red*). The ribosomal RNA and ribosomal proteins are shown in *dark* and *light blue*, respectively. *Top right and middle*: the free, erythromycin- and telithromycin-bound pockets, in D50S (*cyan*) and H50S (*green*), highlighting the differences in sequence and orientation (*green letters in parenthesis refer to the type of the nucleotide in H. marismortui if different from that of D. radiodurans*). *In middle right*, the stacking interactions between telithromycin and the binding pocket in both D50S and H50S are shown by *dotted lines*. Note the superiority of tunnel (pocket) blocking in D50S, compared to mH50S. *Bottom left*: Zoom into D50S macrolide pocket (*cyan*), showing the close proximity between 2058 and the bound erythromycin (*red*). *Bottom right*: Superposition of the locations of three 16-membered macrolide tylosin (*TYL*), carbomycin (*CAR*), and spiramycin (SPI) bound to H50S, on the locations of three 14-membered macrolides, erythromycin (ERY), clarithromycin (CLA), and roxithromycin (ROX) bound to D50S, showing that the 16-membered macrolides should not severely hamper nascent protein passage. The location of A2058 and the approximate tunnel direction are also shown. Note the larger distance between the nucleotide at position 2058 and the desosamine sugars of the three 16-member macrolides, compared to the 14-member compounds.

macrolides with large subunits (Berisio et al., 2003a; Schluenzen et al., 2001; Tu et al., 2005) show that the interactions of the desosamine sugar and the lactone ring play a key role in macrolide binding. These contacts involve predominantly the main constituents of the macrolide-binding pocket, namely nucleotides A2058–A2059 of the 23S RNA Domain V (Figures 5A–5D).

The second macrolide sugar, namely the cladinose, interacts directly with the ribosome only in a few cases (Berisio et al., 2003a; Schluenzen et al., 2001). Three closely related 14-membered macrolides, namely erythromycin and its semisynthetic derivatives, clarithromycin and roxithromycin, exhibit exceptional consistency in their binding modes to the macrolide-binding pocket (Figure 5) (Schluenzen et al., 2001). The high binding affinity of these macrolides was found to originate mainly from their interactions with nucleotide 2058. In all eubacteria this nucleotide is an adenine, which provides the means for prominent macrolide interactions. In eukaryotes, as well as in the archaeon *H. marismortui*, it is a guanine.

Consistently, these structures indicated that owing to increased bulkiness, a guanine in position 2058 should impose spatial constrains and hamper macrolide binding, in accord with the resistance mechanisms that are modifying the chemical identity of this nucleotide either by A → G mutation, or by their methylation (Blondeau et al., 2002; Courvalin et al., 1985; Katz and Ashley, 2005; Poehlsgaard and Douthwaite, 2003; Sigmund et al., 1984; Vester and Douthwaite, 2001; Weisblum, 1995). For over three decades it has been known that mutations in proteins L22 and/or L4 can also induce resistance to the 14-membered antibiotics (Davydova et al., 2002; Pereyre et al., 2002; Poehlsgaard and Douthwaite, 2003; Wittmann et al., 1973). Although these proteins are rather close to the macrolide-binding pocket, the structures of the macrolide complexes do not indicate a direct contact with these proteins. Nevertheless, the increase in A2508 size accompanied with the alterations in the tunnel conformation at its constriction, similar or identical to those seen crystallographically (Berisio et al., 2003a) or by electron microscopy (Gabashvili et al., 2001), could be correlated with this antibiotic-resistant mechanism. Thus, at its swung conformation, the tip of protein L22 hairpin loop, reached protein L4.

To circumvent the acute problems associated with macrolide resistance by modification of A2058, several new compounds have been designed. These include macrolide derivatives, in which the core macrolactone ring has been modified, to 15- (e.g. azithromycin) or 16-(tylosin, carbomycin A, spiramycin, and josamycin) membered rings, all exhibiting activity against some MLSB resistance strains (Bryskier et al., 1993; Poulsen et al., 2000). Ketolides present a yet another chemical approach, based on the addition of rather long extensions, such as alkyl-aryl or quinollyallyl, to the core macrolactone ring, expected to provide additional interactions, thus minimizing the contribution of 2058-9 region.

Drug binding to ribosomes with guanine at position 2058 may superficially indicate low level of selectivity, hence 12 ribosomes, nascent proteins, chaperones, and antibiotics should reduce its clinical relevance. This intriguing question triggered a through comparison between antibiotics binding modes to eubacteria, represented by D50S and to eukaryotes, represented by H50S. This comparison revealed a prominent difference in the effectiveness of tunnel blockage (Figure 5), which could be linked to the specific architecture of the two macrolide-binding pockets.

Indeed, in H50S there are seven nucleotides that differ from the typical eubacteria, among them three present purine/pyrimidine exchange, and most of the conserved nucleotides have different conformations (Figure 5). Accordingly, the binding modes, and consequently, the therapeutical usefulness of macrolides that bind to H50S, namely the 16-membered ring compounds (Hansen et al., 2002b) are considerably different from those found in D50S. Thus, in D50S the macrolides occupy most of the tunnel space, whereas in H50S the 16-membered ring macrolides lie almost parallel to the tunnel wall and consume a smaller part of it (Figure 5). These differences are likely to result from the sequence and conformational divergence of the macrolide-binding pocket, in accordance with the low drug affinity to H50S, which forced the usage of immense excess (several orders of magnitude above the clinical levels) of antibiotics for obtaining measurable binding to H50S (Hansen et al., 2002b, 2003), contrary to the usages of clinically relevant drug concentrations in the complexes of D50S (Auerbach et al., 2004; Bashan and Yonath, 2005; Berisio et al., 2003a, b; Harms et al., 2004; Schluenzen et al., 2001, 2003; Yonath, 2005). Hence, the crystallographically observed differences in the antibiotics binding modes demonstrate the interplay between structure and clinical implications and illuminate the distinction between medically meaningful and less relevant binding.

5.2. G → A MUTATION ENABLES MACROLIDES BINDING

Further comparison, supporting the above conclusions, became possible as G2058 in *H. marismortui* 23S rRNA has recently been mutated to an adenosine (Tu et al., 2005). This mutation (called below mH50S) increases macrolide-binding affinity by 10,000-fold, but did not significantly improve the effectiveness of the binding mode, as the magnitude of tunnel blockage in mH50S remains lower than that achieved by the same drug in the eubacterial D50S (Figures 5). Furthermore, based on azithromycin-binding mode to mH50S (Tu et al., 2005), the impressive gain in drug affinity, achieved by the G2058A mutation, is not accompanied by a comparable alteration in its binding mode compared to H50S wild-type, where 2058 is a guanosine. This seemingly surprising finding indicates that although 2058 identity determines whether binding occurs, the conformations and the chemical identities of the other nucleotide in the macrolide pocket govern the antibiotics-binding modes and, subsequently, the drug effectiveness. Interestingly, all mH50S bound macrolides/ketolides share a similar macrolactone conformation, which is almost identical to that suggested by NMR studies to be of the lowest free energy at ribosome-free environments, therefore more likely to occur in vacuum or dilutes solutions. These experiments ignored the ribosome, which by providing a significant interaction network, alters radically the drug environment. Hence, the preservation of conformation of the drug in isolation is inconsistent with the high-binding affinities between ribosomes and macrolides/ketolides.

The case of telithromycin-mH50S complex (Figure 5) supports the separation between binding and effectiveness. Thus, in mH50S telithromycin does not create the prominent inter-actions of ketolides with domains II (Figure 5), which are consistent with resistance data, and independently identified by footprinting, mutagenesis (e.g. Hansen et al., 1999; Vester and Douthwaite, 2001; Xiong et al., 1999), and crystallographic experiments, using the eubacterium *D. radiodurans* (Berisio et al., 2003a; Schluenzen et al., 2003). Likewise, significant of similarity between the binding modes of telithromycin and erythromycin is inconsistent with the profound differences detected between the susceptibility of A2058G ribosomes to ketolides, as compared with no influence on the susceptibility to macrolides (Pfister et al., 2005).

16. Berisio, R., Harms, J., Schluenzen, F., Zarivach, R., and Hansen, H.A., et al. (2003b). Structural insight into the antibiotic action of telithromycin against resistant mutants. *J. Bacteriol.* 185, 4276–4279.

17. Blaha, G., Wilson, D.N., Stoller, G., Fischer, G., and Willumeit, R., et al. (2003). Localization of the trigger factor binding site on the ribosomal 50S subunit. *J. Mol. Biol.* 326, 887–897.

18. Blondeau, J.M., DeCarolis, E., Metzler, K.L., and Hansen, G.T. (2002). The macrolides. *Expert Opin. Investig. Drugs* 11, 189–215.

19. Bocchetta, M., Xiong, L., and Mankin, A.S. (1998). 23S rRNA positions essential for tRNA binding in ribosomal functional sites. *Proc. Natl. Acad. Sci. USA* 95, 3525–3530.

20. Brodersen, D.E., Clemons, W.M., Jr., Carter, A.P., Morgan-Warren, R.J., and Wimberly, B.T., et al. (2000). The structural basis for the action of the antibiotics tetracycline, pactamycin, and hygromycin B on the 30S ribosomal subunit. *Cell* 103, 1143–1154.

21. Bryskier, A., Butzler, J.P., Neu, H.C., and Tulkens, P.M. (1993). Macrolides-Chemistry, Pharmacology, and Clinical Uses. Oxford.

22. Bukau, B., Deuerling, E., Pfund, C., and Craig, E.A. (2000). Getting newly synthesized proteins into shape. *Cell* 101, 119–122.

23. Cannone, J.J., Subramanian, S., Schnare, M.N., Collett, J.R., and D'Souza, L.M., et al. (2002). The Comparative RNA Web (CRW) Site: an online database of comparative sequence and structure information for ribosomal, intron, and other RNAs. *BMC Bioinformatics* 3, 2.

24. Carter, A.P., Clemons, W.M., Brodersen, D.E., Morgan-Warren, R.J., and Wimberly, B.T., et al. (2000). Functional insights from the structure of the 30S ribosomal subunit and its interactions with antibiotics. *Nature* 407, 340–348.

25. Cooperman, B.S., Wooten, T., Romero, D.P., and Traut, R.R. (1995). Histidine 229 in protein L2 is apparently essential for 50S peptidyl transferase activity. *Biochem. Cell. Biol.* 73, 1087–1094.

26. Courvalin, P., Ounissi, H., and Arthur, M. (1985). Multiplicity of macrolide-lincosamide-streptogramin antibiotic resistance determinants. *J. Antimicrob. Chemother.* 16, 91–100.

27. Davydova, N., Streltsov, V., Wilce, M., Liljas, A., and Garber, M. (2002). L22 ribosomal protein and effect of its mutation on ribosome resistance to erythromycin. *J. Mol. Biol.* 322, 635–644.

28. Deuerling, E., Schulze-Specking, A., Tomoyasu, T., Mogk, A., and Bukau, B. (1999). Trigger factor and DnaK cooperate in folding of newly synthesized proteins. *Nature* 400, 693–696.

29. Dorner, S., Polacek, N., Schulmeister, U., Panuschka, C., and Barta, A. (2002). Molecular aspects of the ribosomal peptidyl transferase. *Biochem. Soc. Trans.* 30, 1131–1136.

30. Eisenstein, M., Hardesty, B., Odom, O.W., Kudlicki, W., and Kramer, G., et al. (1994). *Modeling and experimental study of the progression of nascent protein in ribosomes; in Supramolecular Structure and Function*, Pifat, G. (ed.), pp. 213–246, Balaban Press, Rehovot, Israel.

31. Etchells, S.A. and Hartl, F.U. (2004). The dynamic tunnel. *Nat. Struct. Mol. Biol.* 11, 391–392.

32. Ferbitz, L., Maier, T., Patzelt, H., Bukau, B., and Deuerling, E., et al. (2004). Trigger factor in complex with the ribosome forms a molecular cradle for nascent proteins. *Nature* 431, 590–596.

33. Frydman, J., (2001). Folding of newly translated proteins in vivo: the role of molecular chaperones. *Annu. Rev. Biochem.* 70, 603–647.

34. Gabashvili, I.S., Gregory, S.T., Valle, M., Grassucci, R., and Worbs, M., et al. (2001). The polypeptide tunnel system in the ribosome and its gating in erythromycin resistance mutants of L4 and L22. *Mol. Cell* 8, 181–188.

35. Gale, E.F., Cundliffe, E., Reynolds, P.E., Richmond, M.H., and Waring, M.J. (1981). The Molecular Basis of Antibiotic Action, Wiley, London, pp. 419–439.

36. Gaynor, M. and Mankin, A.S. (2003). Macrolide antibiotics: binding site, mechanism of action, resistance. *Curr. Top Med. Chem.* 3, 949–961.

37. Gilbert, R.J., Fucini, P., Connell, S., Fuller, S.D., and Nierhaus, K.H., et al. (2004). Three-dimensional structures of translating ribosomes by cryo-eM. *Mol. Cell* 14, 57–66.

38. Goldberg, I.H. and Mitsugi, K. (1966). Sparsomycin, an inhibitor of aminoacyl transfer to polypeptide. *Biochem. Biophys. Res. Commun.* 23, 453–459.

39. Gong, F. and Yanofsky, C. (2002). Instruction of translating ribosome by nascent peptide. *Science* 297, 1864–1867.

40. Gottesman, M.E. and Hendrickson, W.A. (2000). Protein folding and unfolding by *Escherichia coli* chaperones and chaperonins. *Curr. Opin. Microbiol.* 3, 197–202.

41. Green, R., Samaha, R.R., and Noller, H.F. (1997). Mutations at nucleotides G2251 and U2585 of 23 S rRNA perturb the peptidyl transferase center of the ribosome. *J. Mol. Biol.* 266, 40–50.

42. Gregory, S.T. and Dahlberg, A.E. (2004). Peptide bond formation is all about proximity. *Nat. Struct. Mol. Biol.* 11, 586–587.

43. Gregory, S.T., Carr, J.F., Rodriguez-Correa, D., and Dahlberg, A.E. (2005). Mutational analysis of 16S and 23S rRNA genes of *Thermus thermophilus. J. Bacteriol.* 187, 4804–4812.

44. Hansen, L.H., Mauvais, P., and Douthwaite, S. (1999). The macrolide-ketolide antibiotic binding site is formed by structures in domains II and V of 23S ribosomal RNA. *Mol. Microbiol.* 31, 623–631.

45. Hansen, J.L., Schmeing, T.M., Moore, P.B., and Steitz, T.A. (2002a). Structural insights into peptide bond formation. *Proc. Natl. Acad. Sci. USA* 99, 11670–11675.

46. Hansen, J.L., Ippolito, J.A., Ban, N., Nissen, P., and Moore, P.B., et al. (2002b). The structures of four macrolide antibiotics bound to the large ribosomal subunit. *Mol. Cell* 10, 117–128.

47. Hansen, J.L., Moore, P.B., and Steitz, T.A. (2003). Structures of five antibiotics bound at the peptidyl transferase center of the large ribosomal subunit. *J. Mol. Biol.* 330, 1061–1075.

48. Hardesty, B., Kudlicki, W., Odom, O.W., Zhang, T., and McCarthy, D., et al. (1995). Cotranslational folding of nascent proteins on *Escherichia coli* ribosomes. *Biochem. Cell. Biol.* 73, 1199–1207.

49. Harms, J., Schluenzen, F., Zarivach, R., Bashan, A., and Gat, S., et al. (2001). High resolution structure of the large ribosomal subunit from a mesophilic eubacterium. *Cell* 107, 679–688.

50. Harms, J., Schluenzen, F., Fucini, P., Bartels, H., and Yonath, A. (2004). Alterations at the peptidyl transferase centre of the ribosome induced by the synergistic action of the streptogramins dalfopristin and quinupristin. *BMC Biol.* 2, 1–10.

51. Hartl, F.U. and Hayer-Hartl, M. (2002). Molecular chaperones in the cytosol: from nascent chain to folded protein. *Science* 295, 1852–1858.

52. Jencks, W.P. (1969, reissued 1987). Catalysis in Chemistry and Enzymology. Dover Publications, McGraw-Hill, Mineola, NY.

53. Johnson, A.E. (2005). The co-translational folding and interactions of nascent protein chains: a new approach using fluorescence resonance energy transfer. *FEBS Lett.* 579, 916–920.

54. Katz, L. and Ashley, G.W. (2005). Translation and protein synthesis: macrolides. *Chem. Rev.* 105, 499–528.

55. Kim, D.F. and Green, R. (1999). Base-pairing between 23S rRNA and tRNA in the ribosomal A site. *Mol. Cell* 4, 859–864.

56. Knowles, D.J., Foloppe, N., Matassova, N.B., and Murchie, A.I. (2002). The bacterial ribosome, a promising focus for structurebased drug design. *Curr. Opin. Pharmacol.* 2, 501–506.

57. Kramer, G., Rauch, T., Rist, W., Vorderwulbecke, S., and Patzelt, H., et al. (2002). L23 protein functions as a chaperone docking site on the ribosome. *Nature* 419, 171–174.

58. Kristensen, O. and Gajhede, M. (2003). Chaperone binding at the ribosomal exit tunnel. *Structure* 11, 1547–1556.

59. Ludlam, A.V., Moore, B.A., and Xu, Z. (2004). The crystal structure of ribosomal chaperone trigger factor from *Vibrio cholerae. Proc. Natl. Acad. Sci. USA* 101, 13436–13441.

60. Maier, T., Ferbitz, L., Deuerling, E., and Ban, N. (2005). A cradle for new proteins: trigger factor at the ribosome. *Curr. Opin. Struct. Biol.* 15, 204–212.

61. Mankin, A.S. and Garrett, R.A. (1991). Chloramphenicol resistance mutations in the single 23S rRNA gene of the archaeon *Halobacterium halobium. J. Bacteriol.* 173, 3559–3563.

62. Milligan, R.A. and Unwin, P.N. (1986). Location of exit channel for nascent protein in 80S ribosome. *Nature* 319, 693–695.

63. Monro, R.E., Celma, M.L., and Vazquez, D. (1969). Action of sparsomycin on ribosome-catalysed peptidyl transfer. *Nature* 222, 356–358.
64. Moore, P.B. and Steitz, T.A. (2003). After the ribosome structures: how does peptidyl transferase work? *RNA* 9, 155–159.
65. Nakatogawa, H. and Ito, K. (2002). The ribosomal exit tunnel functions as a discriminating gate. *Cell* 108, 629–636.
66. Nierhaus, K.H., Schulze, H., and Cooperman, B.S. (1980). Molecular mechanisms of the ribosomal peptidyl transferase center. *Biochem. Int.* 1, 185–192.
67. Nissen, P., Hansen, J., Ban, N., Moore, P.B., and Steitz, T.A. (2000). The structural basis of ribosome activity in peptide bond synthesis. *Science* 289, 920–930.
68. Noller, H.F., Hoffarth, V., and Zimniak, L. (1992). Unusual resistance of peptidyl transferase to protein extraction procedures. *Science* 256, 1416–1419.
69. Pereyre, S., Gonzalez, P., De Barbeyrac, B., Darnige, A., and Renaudin, H., et al. (2002). Mutations in 23S rRNA account for intrinsic resistance to macrolides in *Mycoplasma hominis* and *Mycoplasma fermentans* and for acquired resistance to macrolides in *M. hominis*. *Antimicrob Agents Chemother.* 46, 3142–3150.
70. Pfister, P., Jenni, S., Poehlsgaard, J., Thomas, A., and Douthwaite, S., et al. (2004). The structural basis of macrolide-ribosome binding assessed using mutagenesis of 23S rRNA positions 2058 and 2059. *J. Mol. Biol.* 342, 1569–1581.
71. Pfister, P., Corti, N., Hobbie, S., Bruell, C., and Zarivach, R., et al. (2005). 23S rRNA base pair 2057-2611 determines ketolide susceptibility and fitness cost of the macrolide resistance mutation 2058A → G. *Proc. Natl. Acad. Sci. USA* 102, 5180–5185.
72. Pioletti, M., Schluenzen, F., Harms, J., Zarivach, R., and Gluehmann, M., et al. (2001). Crystal structures of complexes of the small ribosomal subunit with tetracycline, edeine and IF3. *EMBO J.* 20, 1829–1839.
73. Poehlsgaard, J. and Douthwaite, S. (2003). Macrolide antibiotic interaction and resistance on the bacterial ribosome. *Curr. Opin. Invest. Drugs* 4, 140–148.
74. Polacek, N., Gomez, M.J., Ito, K., Xiong, L., and Nakamura, Y., et al. (2003). The critical role of the universally conserved A2602 of 23S ribosomal RNA in the release of the nascent peptide during translation termination. *Mol. Cell* 11, 103–112.
75. Porse, B.T., Kirillov, S.V., Awayez, M.J., Ottenheijm, H.C., and Garrett, R.A. (1999). Direct crosslinking of the antitumor antibiotic sparsomycin, and its derivatives, to A2602 in the peptidyl transferase center of 23S-like rRNA within ribosome – tRNA complexes. *Proc. Natl. Acad. Sci. USA* 96, 9003–9008.
76. Poulsen, S.M., Kofoed, C., and Vester, B. (2000). Inhibition of the ribosomal peptidyl transferase reaction by the mycarose moiety of the antibiotics carbomycin, spiramycin and tylosin. *J. Mol. Biol.* 304, 471–481.
77. Rospert, S. (2004). Ribosome function: governing the fate of a nascent polypeptide. *Curr. Biol.* 14, R386–388.
78. Sanz, J.L., Marin, I.R.A., and Urena, D. (1993). Functional analysis of seven ribosomal systems from extreme halophilic archaea. *Can. J. Microbiol.* 35, 311–317.
79. Schluenzen, F., Tocilj, A., Zarivach, R., Harms, J., and Gluehmann, M., et al. (2000). Structure of functionally activated small ribosomal subunit at 3.3 angstroms resolution. *Cell* 102, 615–623.
80. Schluenzen, F., Zarivach, R., Harms, J., Bashan, A., and Tocilj, A., et al. (2001). Structural basis for the interaction of antibiotics with the peptidyl transferase centre in eubacteria. *Nature* 413, 814–821.
81. Schluenzen, F., Harms, J.M., Franceschi, F., Hansen, H.A., and Bartels, H., et al. (2003). Structural basis for the antibiotic activity of ketolides and azalides. *Structure* 11, 329–338.

82. Schluenzen, F., Pyetan, E., Fucini, P., Yonath, A., and Harms, J. (2004). Inhibition of peptide bond formation by pleuromutilins: the structure of the 50S ribosomal subunit from *Deinococcus radiodurans* in complex with tiamulin. *Mol. Microbiol.* 54, 1287–1294.

83. Schmeing, T.M., Seila, A.C., Hansen, J.L., Freeborn, B., and Soukup, J.K., et al. (2002). A pre-translocational intermediate in protein synthesis observed in crystals of enzymatically active 50S subunits. *Nat. Struct. Biol.* 9, 225–230.

84. Schmeing, T.M., Moore, P.B., and Steitz, T.A. (2003). Structures of deacylated tRNA mimics bound to the E site of the large ribosomal subunit. *RNA* 9, 1345–1352.

85. Schmeing, T.M., Huang, K.S., Kitchen D.E., Strobel, S.A., and Steitz T.A. (2005a). Structural insights into the roles of water and the 20 hydroxyl of the P Site tRNA in the peptidyl transferase Reaction. *Mol. Cell* 20, 437–448

86. Schmeing, T.M., Huang, K.S., Strobel, S.A., and Steitz T.A. (2005b). An induced-fit mechanism to promote peptide bond formation and exclude hydrolysis of peptidyl-tRNA, *Nature* 438, 520–525

87. Shevack, A., Gewitz, H.S., Hennemann, B., Yonath, A., and Wittmann, H.G. (1985). Characterization and crystallization of ribosomal particles from *Halobacterium marismortui*. *FEBS Lett.* 184, 68–71.

88. Sievers, A., Beringer, M., Rodnina, M.V., and Wolfenden, R. (2004). The ribosome as an entropy trap. *Proc. Natl. Acad. Sci. USA* 101, 7897–7901.

89. Sigmund, C.D., Ettayebi, M., and Morgan, E.A. (1984). Antibiotic resistance mutations in 16S and 23S ribosomal RNA genes of *Escherichia coli*. *Nucleic Acids Res.* 12, 4653–4663.

90. Spahn, C.M. and Prescott, C.D. (1996). Throwing a spanner in the works: antibiotics and the translation apparatus. *J. Mol. Med.* 74, 423–439.

91. Stroud, R.M. and Walter, P. (1999). Signal sequence recognition and protein targeting. *Curr. Opin. Struct. Biol.* 9, 754–759.

92. Schuwirth, B.S., Borovinskaya, M.A., Hau,C.W., Zhang, W., Vila-Sanjurjo, A., Holton, J.M., and Cate, J. (2005). Structures of the bacterial ribosome at 3.5 a resolution. *Science* 310, 827-34.

93. Tan, G.T., DeBlasio, A., and Mankin, A.S. (1996). Mutations in the peptidyl transferase center of 23 S rRNA reveal the site of action of sparsomycin, a universal inhibitor of translation. *J. Mol. Biol.* 261, 222–230.

94. Tenson, T. and Ehrenberg, M. (2002). Regulatory nascent peptides in the ribosomal tunnel. *Cell* 108, 591–594.

95. Thirumalai, D. and Lorimer, G.H. (2001). Chaperonin-mediated protein folding. *Annu. Rev. Biophys. Biomol. Struct.* 30, 245–269.

96. Thompson, J. and Dahlberg, A.E. (2004). Testing the conservation of the translational machinery over evolution in diverse environments: assaying *Thermus thermophilus* ribosomes and initiation factors in a coupled transcription-translation system from *Escherichia coli*. *Nucleic Acids Res.* 32, 5954–5961.

97. Thompson, J., Kim, D. F., O'Connor, M., Lieberman, K. R., Bayfield, M. A., et al. (2001). Analysis of mutations at residues A2451 and G2447 of 23S rRNA in the peptidyltransferase active site of the 50S ribosomal subunit. *Proc. Natl. Acad. Sci. USA* 98, 9002–9007.

98. Tu, D., Blaha, G., Moore, P.B., and Steitz, T.A. (2005). Structures of MLSBK antibiotics bound to mutated large ribosomal subunits provide a structural explanation for resistance. *Cell* 121, 257–270.

99. Unge, J., berg, A., Al-Kharadaghi, S., Nikulin, A., and Nikonov, S., et al. (1998). The crystal structure of ribosomal protein L22 from *Thermus thermophilus*: insights into the mechanism of erythromycin resistance. *Structure* 6, 1577–1586.

100. Vazquez, D. (1979). Inhibitors of protein biosynthesis. *Mol. Biol. Biochem. Biophys.* 30, 1–312.

101. Vester, B. and Douthwaite, S. (2001). Macrolide resistance conferred by base substitutions in 23S rRNA. *Antimicrob Agents Chemother.* 45, 1–12.

102. Walter, P. and Johnson, A.E. (1994). Signal sequence recognition and protein targeting to the endoplasmic reticulum membrane. *Annu. Rev. Cell. Biol.* 10, 87–119.

103. Weinger, J.S., Parnell, K.M., Dorner, S., Green, R., and Strobel, S.A. (2004). Substrate-assisted catalysis of peptide bond formation by the ribosome. *Nat. Struct. Mol. Biol.* 11, 1101–1106.

104. Weisblum, B. (1995). Erythromycin resistance by ribosome modification. *Antimicrob Agents Chemother.* 39, 577–585.

105. White, S.H. and von Heijne, G. (2004). The machinery of membrane protein assembly. *Curr. Opin. Struct. Biol.* 14, 397–404.

106. Wimberly, B.T., Brodersen, D.E., Clemons, W.M., Jr., Morgan-Warren, R.J., and Carter, A.P., et al. (2000). Structure of the 30S ribosomal subunit. *Nature* 407, 327–339.

107. Wittmann, H.G., Stoffler, G., Apirion, D., Rosen, L., and Tanaka, K., et al. (1973). Biochemical and genetic studies on two different types of erythromycin resistant mutants of *Escherichia coli* with altered ribosomal proteins. *Mol. Gen. Genet.* 127, 175–189.

108. Woolhead, C.A., McCormick, P.J., and Johnson, A.E. (2004). Nascent membrane and secretory proteins differ in FRET detected folding far inside the ribosome and in their exposure to ribosomal proteins. *Cell* 116, 725–736.

109. Xiong, L., Shah, S., Mauvais, P., and Mankin, A.S. (1999). A ketolide resistance mutation in domain II of 23S rRNA reveals the proximity of hairpin 35 to the peptidyl transferase centre. *Mol. Microbiol.* 31, 633–639.

110. Yonath, A. (2002). The search and its outcome: high-resolution structures of ribosomal particles from mesophilic, thermophilic, and halophilic bacteria at various functional states. *Annu. Rev. Biophys. Biomol. Struct.* 31, 257–273.

111. Yonath, A. (2003a). Structural insight into functional aspects of ribosomal RNA targeting. *Chem. Biol. Chemistry* 4, 1008–1017.

112. Yonath, A. (2003b). Ribosomal tolerance and peptide bond formation. *Biol. Chem.* 384, 1411–9.

113. Yonath, A. (2005). Antibiotics targeting ribosomes: resistance, selectivity, synergism, and cellular regulation. *Annu. Rev. Biochem.* 74, 649–679.

114. Yonath, A. and Bashan, A. (2004). Ribosomal crystallography: initiation, peptide bond formation, and amino acid polymerization are hampered by antibiotics. *Annu. Rev. Microbiol.* 58, 233–251.

115. Yonath, A., Leonard, K. R., and Wittmann, H. G. (1987). A tunnel in the large ribosomal subunit revealed by three-dimensional image reconstruction. *Science* 236, 813–816.

116. Youngman, E.M., Brunelle, J.L., Kochaniak, A.B., and Green, R. (2004). The active site of the ribosome is composed of two layers of conserved nucleotides with distinct roles in peptide bond formation and peptide release. *Cell* 117, 589–599.

117. Yusupov, M.M., Yusupova, G.Z., Baucom, A., Lieberman, K., and Earnest, T.N., et al. (2001). Crystal structure of the ribosome at 5.5 A resolution. *Science* 292, 883–896.

118. Zarivach, R., Bashan, A., Berisio, R., Harms, J., and Auerbach, T., et al. (2004). Functional aspects of ribosomal architecture: symmetry, chirality and regulation. *J. Phys. Org. Chem.* 17, 901–912.

AUTHOR INDEX